Oxford International Primary Science

Workbook

5

Terry Hudson
Debbie Roberts

Great Clarendon Street, Oxford, OX2 6DP, United Kingdom

Oxford University Press is a department of the University of Oxford. It furthers the University's objective of excellence in research, scholarship, and education by publishing worldwide. Oxford is a registered trade mark of Oxford University Press in the UK and in certain other countries

© Terry Hudson, Debbie Roberts 2016

The moral rights of the authors have been asserted

First published in 2016

All rights reserved. No part of this publication may be reproduced, stored in a retrieval system, or transmitted, in any form or by any means, without the prior permission in writing of Oxford University Press, or as expressly permitted by law, by licence or under terms agreed with the appropriate reprographics rights organization. Enquiries concerning reproduction outside the scope of the above should be sent to the Rights Department, Oxford University Press, at the address above.

You must not circulate this work in any other form and you must impose this same condition on any acquirer

British Library Cataloguing in Publication Data
Data available

9780198376460

10

Paper used in the production of this book is a natural, recyclable product made from wood grown in sustainable forests. The manufacturing process conforms to the environmental regulations of the country of origin.

Printed by Repro India Ltd.

Acknowledgements
The publishers would like to thank the following for permissions to use their photographs:

Cover: Paul Souders/Corbis
p12: Classic Image/Alamy Stock Photo; **p15:** Bigstock.
All other photos by Shutterstock.

Although we have made every effort to trace and contact all copyright holders before publication this has not been possible in all c ases. If notified, the publisher will rectify any errors or omissions at the earliest opportunity.

Links to third party websites are provided by Oxford in good faith and for information only. Oxford disclaims any responsibility for the materials contained in any third party website referenced in this work.

Contents

How to use this book	4
Module 1 The Way We See Things	6
Module 2 Evaporation and Condensation	20
Module 3 The Life Cycle of a Flowering Plant	34
Module 4 Investigating Plant Growth	48
Module 5 Earth's Movements	62
Module 6 Shadows	76
Quiz yourself	90

How to use this book

This workbook has been written to support the Student Book that students are using at school. The Student Book has some write-in and hands-on tasks to help students to learn and test their understanding, but it is important to extend these tasks – including home learning.

This workbook is split into the six modules of the Cambridge curriculum for this stage. These are:

1 The Way We See Things
2 Evaporation and Condensation
3 The Life Cycle of a Flowering Plant
4 Investigating Plant Growth
5 Earth's Movements
6 Shadows

Each module begins with extra support to enable you to help the student. It explains the focus of the module and then provides specific information and advice about how the student can be supported. We all learn new skills and knowledge initially with lots of support but we cannot continue to learn if we always have that support. In other words, allow the student to try things on their own or with some guidance, and only step in if the student shows genuine confusion or frustration.

The activities build on the work at school and are aimed at developing language skills, scientific enquiry skills and understanding rather than just recall.

Each module ends with a review task that the student is asked to do at home. This review and reflection is a key aspect of learning.

Why is home learning important?

Encouraging students to think about and apply their growing skills and knowledge outside the classroom, and especially at home, allows them to consolidate understanding and to practise activities. This helps with confidence. They also have opportunities to see that science is relevant all around them – not only during science activities in school. Another advantage of home learning is that you can find out what the student is studying and show your genuine interest. The student may even be able to teach you some science. Finally, home learning can be fun and help the student to develop good learning and study habits that will help them throughout their life.

Class activities and home learning activities

Each activity in this workbook has an icon to give the student, you and the student's teachers a clear idea of the nature of the task. These icons are explained on page 5.

Each module has four **class activities** and eight **home learning activities**.

- Teachers will use the class activities to supplement the Student Book and additional resources. The class activities involve group discussion with peers and activities requiring equipment not easily obtained in the home.

- The home learning activities cover a range of different activity types that encourage the development of science skills. In many of the home learning tasks the student is asked to involve people at home. For example, the student asks the people at home to take part in team games, discussions, investigations, surveys and research tasks, and to help the student make objects such as a periscope and silhouettes.

Activities and icons

Writing activities
There are spaces for students to record short answers to key questions using the information on the pages and from their prior learning in school. Sometimes a drawing or graph is required.

Discussion activities
Students are encouraged to discuss scientific ideas and approaches and are expected to work in pairs and small groups for this type of work.

Investigations
Students are encouraged to record plans, scientific notes and results for each investigation. They are asked to make predictions and to compare their results with others.

Measuring activities
Students are given opportunities to use and develop a number of key measuring skills. There are step-by-step instructions and advice about how to use measuring equipment accurately.

'Think about' questions
Students are encouraged to consider scientific phenomena and to try to explain them using the knowledge and understanding they have gained during the topic. These questions may also involve thinking of examples from students' own experience.

Extra support
This icon indicates where the adult at home can find advice on how to help the student with each home learning activity.

1 The Way We See Things

Extra support

Introduction

The main purpose of this module is to help students to understand more about light. They learn about the need for light in order to see things and they discuss some sources of light. The first two activities help students to realise that some objects produce light and others simply reflect it.

The next activities are designed to help students to understand that shadows occur when light is blocked. They learn that some materials are good at blocking light (they are opaque) and others are not (they are transparent).

Students explore the use of mirrors and shiny surfaces to reflect light. They study mirror images: the image remains the same way up as the original object but left becomes right and right becomes left. They also examine some uses of reflected light.

This module will help students to practise these scientific enquiry skills:

- planning – asking questions and planning how to seek answers in a fair test (pages 8, 10, 14, 17, 18)
- predicting – stating what they think will happen and then comparing this with what actually happens (page 8)
- observation – collecting sufficient evidence, understanding the need for repeated observations or measurements (pages 8, 9, 10, 11, 13, 15, 16, 17, 18)
- recording – writing or drawing observations or stages in work and presenting results in line graphs or bar charts (pages 8, 9, 10, 11, 12, 13, 15, 16, 17, 18)
- making comparisons – comparing sets of evidence or data (pages 8, 9, 10, 18)
- drawing conclusions – examining results to identify any patterns and/or to suggest explanations (pages 10, 16, 17, 18)
- evaluating results – evaluating the investigation and results and suggesting ways to improve the investigation (pages 10, 14).

Ways to help

It is important to emphasise that we see all objects because light is reflected from them and that shiny or mirrored surfaces are particularly good at reflecting light.

Allow the student to walk towards a light source, then turn and see how their shadow has changed. Showing the shadow of a window is useful. The glass does not cast a shadow but the opaque frame does.

Remember to encourage and help the student but try to let them find out as much as they can on their own.

Finally, help the student to complete the 'What I have learned…' summary to test their understanding and recall.

> **Key words**
> beam   object   shadow
> light   opaque   translucent
> light source   ray   transparent
> mirror   reflect/reflection

> **Scientific enquiry words**
> explain   observe   record
> investigate   plan   show
> measure   predict

Helping with activities

The following guidance is intended to offer advice to the parent, or other adult at home, on how to help the student with each home learning activity.

Reflection at home (page 9)

Help the student to find mirrors at home. Help them to work out what each mirror is used for – you could include design uses such as brightening up a dark area of a house or making a room look bigger. If you go out shopping or visiting encourage the student to look for examples of mirrors in use in the local area.

Seeing through materials (page 11)

Support the student in learning the technical terms 'opaque', 'translucent' and 'transparent' by encouraging them to write the words and display them. Let them survey their home to find examples of materials with each of these properties, and help them to see that the property is closely linked to the use of a material.

Mirror writing (page 12)

It is always fun and challenging to try to write while looking at the page in a mirror but it is even harder to read mirror writing without a mirror. Encourage members of the family to guess what the student's sentence says and then use a mirror to decipher it. You could all join in and try writing messages. Point out that in a mirror objects are the correct way up but the image is reversed. The top is still top but right is left.

Your home periscope (page 14)

Help the student to find appropriate materials for the periscope. Any box, preferably narrow, will work. Cut flaps so that two small mirrors can be set at 45°. If you do not have mirrors, old CDs are very reflective and you can cut them carefully to shape. Safety – do not allow the student to cut CDs. Wrap the CD in a cloth while you are cutting it to avoid sharp edges flying off.

Mirrors and design (page 15)

Encourage the student to find pictures of mirrors and choose one of these pictures to draw a ray diagram. Ray diagrams will be an increasingly important aspect of the student's work on light and reflection in future years.

Does light travel in straight lines? (page 17)

The student can demonstrate that light travels in straight lines by using cards with a hole made in the centre of each. Allow them to shine a torch through a card and see if the light passes through. Make sure they realise that unless the holes of the three cards are lined up, the light will not pass through them all.

Investigate colour (page 18)

Help the student by providing a bright torch and some materials to use as coloured filters. Cellophane wrappers (from flowers or confectionary) are ideal but coloured tissue paper and some fabrics also work well. Encourage the student to compare the colour of the objects when viewed using the different coloured light sources.

What I have learned… (page 19)

This summary activity tests the student's understanding and recall from the whole module. Help the student to solve the clues and complete the crossword.

How do our eyes see things?

See Student Book 5, pages 6–7

Class activity Shining back

 Which materials are good and poor reflectors?

Your teacher will make the room darker. You are going to shine a torch onto six different **objects**.

Do not shine a torch directly at someone else. The bright **light** can damage their eyes.

1 Predict what you think the objects will look like in the torchlight.

Will the objects be dull or shiny? Will any of the objects shine the light back? This is called **reflection**.

2 Observe what happens and **record** your results in the table.

Name of the object	What material is the object made from?	Is the object shiny or dull in torchlight?	Is the object a good reflector or poor reflector?

 Which materials were dull? _____

Which materials were shiny? _____

Did any of the materials **reflect** the light? _____

How accurate were your predictions?

How do our eyes see things?

See Student Book 5, pages 6–7

Home learning Reflection at home

We use **mirrors** for many things.

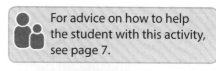
For advice on how to help the student with this activity, see page 7.

A

B

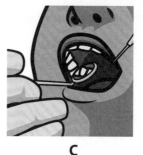
C

What are the mirrors in the pictures being used for?

Picture A _____   Picture B _____   Picture C _____

Carry out a survey of mirrors at home.

1 Find as many mirrors as you can. Note whether they are large, medium-sized or small, and what room they are in.

2 Think about what each mirror is used for. For example, some mirrors are used to make a room look lighter or bigger.

3 **Record** your results in the table.

Size of mirror	Which room is it in?	What is it used for?

What were most of the mirrors used for? _____

Explain why you cannot use a mirror in the dark. _____

How are shadows made?

See Student Book 5, pages 8–9

Class activity Making shadows

 Make a sunshade for a plant.

The plant needs sunlight but not too much. You are going to **investigate** some different materials to see which is best to make a small sunshade for a plant.

Your teacher will give you different materials to test.

1 **Plan** how you will compare the materials.

 What will you use as a **light source**? _____

 How will you **measure** the **shadows**? _____

2 Carry out your test.

3 Design and complete a results table.

 Which material did you choose? Why?

_____

_____

 How could you improve your investigation if you did it again?

How are shadows made?

See Student Book 5, pages 8–9

Home learning Seeing through materials

For advice on how to help the student with this activity, see page 7.

Which parts of these glasses are see-through? _____

Would the glasses be useful if these parts were not see-through?
Yes   No

A see-through material is **transparent**.

Find three examples of transparent objects at home.

Draw the objects in the boxes.

Object 1	Object 2	Object 3

Could any of the objects be made from an **opaque** (non-transparent) material?

Which ones? _____

Which of the objects must be transparent to be useful?
_____

Why must a car windscreen be transparent? Write a sentence to **explain** this.
_____

Some materials allow some but not all light through. These materials are **translucent**.

Name two objects in your home that are made from a translucent material.

1 _____   2 _____

The journey of light

See Student Book 5, pages 10–11

Home learning Mirror writing

Try to read this sentence.

> **Mirrors reflect light.** *(shown in mirror writing)*

Write what the sentence says. _____

Now hold the sentence up to a mirror. Were you correct?   Yes   No

✏️ Use a mirror to create your own secret writing. Write a sentence to describe some science you have done at school. Look in the mirror as you write so that your writing is mirror writing.

My mirror writing

For advice on how to help the student with this activity, see page 7.

The famous scientist, Leonardo da Vinci, wrote his notes using a mirror.

My mirror drawing

💬 Ask people at home whether they can read your mirror writing.

Then give them a mirror so they can find out what you have learned.

✏️ Draw an object at home while looking at it in a mirror.

✏️ How is the real object different from its image in the mirror?

The journey of light

See Student Book 5, pages 12–13

Class activity Make a periscope

 Periscope

Use the diagram to help you make your periscope.

> You will need: two small mirrors, a shoe box or other narrow box made of thin card, scissors and sticky tape.

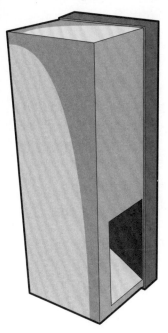
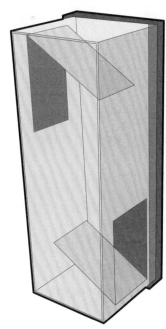

> You can use pieces from an old CD if you do not have any small mirrors.
> Ask your teacher to cut the CDs. Be careful – mirrors and cut CDs have sharp edges.

1. Cut flaps so that two small mirrors can be placed inside the box.
2. Stick the mirrors inside the box, fixing them at a 45° angle.

 Test your periscope.

 Can you see around corners?   Yes   No

Can you see over objects?   Yes   No

 Why do the mirrors have to be at a 45° angle?

The journey of light

See Student Book 5, pages 12–13

Home learning Your home periscope

💬 Show the periscope you made in class to people at home.

Demonstrate how it works.

For advice on how to help the student with this activity, see page 7.

✋ Make a periscope with the people at home.

1 Make your own design. You could use one of the ideas in the pictures.

> You could use one of these to make your periscope:
> - the card tube from a roll of paper towel or wrapping paper
> - the box from clingfilm or aluminium foil
> - a drink carton
> - a shoe box
> - a cereal box.
>
> You will also need two small mirrors, scissors and sticky tape.

> Remember: you can use old CDs for mirrors.
>
> You must let an adult cut the CDs. They can shatter and pieces can go into your eye.
>
> Be very careful of sharp edges.

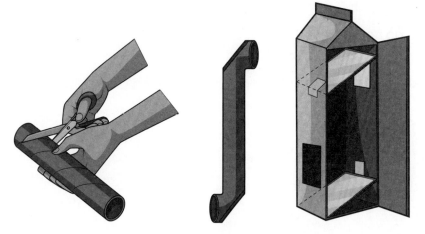

2 **Explain** to your family how the periscope works.

3 Find some objects to look over and around.

💬 How could you make your design better?

The journey of light

See Student Book 5, pages 14–15

Home learning Mirrors and design

Mirrors are used for making the insides and outsides of houses and buildings look brighter and larger. Inside buildings this is a part of the interior design.

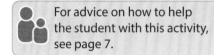
For advice on how to help the student with this activity, see page 7.

 Mirrors used for interior design

1. Look around your home. Are any of the mirrors helping with the interior design of your home?

2. Use magazines or the Internet to find examples of mirrors being used for interior design.

✏️ Draw or cut out and stick an example in the box.

Find the main source of light in the picture.

✏️ Draw a **ray** diagram onto the picture to **show** how the light travels from the light source and where it travels.

✏️ **Explain** how the mirror makes the room brighter.

_____

_____

The journey of light

See Student Book 5, pages 14–15

Class activity Make a pinhole camera

 Make your own camera.

The pinhole camera is one of the earliest types of camera.

Make your camera

1. Find a cardboard box such as a shoe box.
2. Cut out a rectangle from one of the smaller sides.
3. Glue or tape a strip of thin tissue paper over the hole to make a screen.
4. Make a tiny 'pinhole' in the middle of the side opposite your screen.

   Make this hole as small as possible. You can make it a bit bigger later if you need to.
5. Place the top of the box back on to keep the light out of your camera.

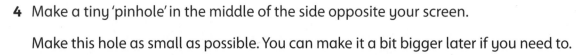

Use your camera

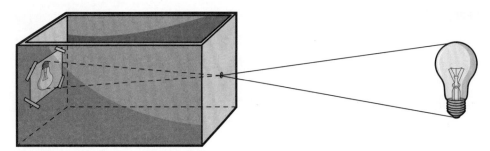

1. Point the hole at a light source such as a light bulb or a candle.

   If you cannot see anything on the screen, make your pinhole slightly bigger.

   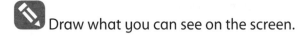 Draw what you can see on the screen.

2. Take your camera outside and point it at bright objects.

    What do you notice about the image of the objects?

Do not point your camera at the Sun.

Light changing direction

See Student Book 5, pages 16–17

Home learning Does light travel in straight lines?

 How does light travel?

 For advice on how to help the student with this activity, see page 7.

Use three pieces of card and a torch or light bulb to **investigate** how light travels.

1 Make a small hole in the centre of each piece of card.

   Use a hole punch or carefully push a pencil through.

   Make sure the hole is in the same position on all the cards.

2 Stand the cards up using modelling clay or reusable adhesive.

You will need: three pieces of thick card, a hole punch or pencil, modelling clay or reusable adhesive, a torch or lit bulb and a small area of blank wall or a large piece of card covered with paper.

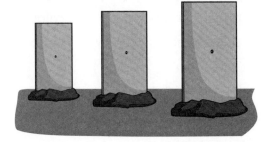

3 Shine a torch or lit bulb onto the cards.

4 **Investigate** how to arrange the cards so that the light passes through all the holes and shines onto the wall or a paper screen behind.

 How did you have to line up the holes? _____

Does this **show** that light travels in a straight line? **Explain** your answer.

_____

 Draw your apparatus. **Show** the light as a **beam** leaving the light source and hitting the wall or paper screen.

Magnificent colours

See Student Book 5, pages 18–19

Home learning Investigate colour

How does coloured light change the colour of objects?

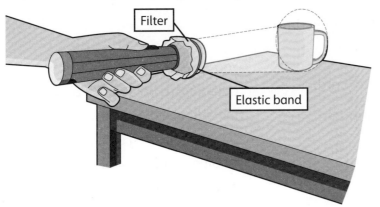

 For advice on how to help the student with this activity, see page 7.

You will need: a torch, four coloured filters, an elastic band and four different objects.

For the filters you can use any coloured material that allows light through, for example coloured tissue paper, lightweight fabric or coloured cellophane, such as sweet wrappers.

1 Shine your torch through each filter.

What happens to the colour of the torch beam? _____

2 Shine different coloured light onto some objects in a darkened room.

3 **Observe** and **record** the colour of each object in daylight.

4 **Observe** and **record** the colour of each object with the different filtered light.

5 Complete the table.

Object	Colour in daylight	Colour with a _____ filter	Colour with a _____ filter	Colour with a _____ filter	Colour with a _____ filter

What happened to the colour of the objects in different coloured light? _____

Explain why this happens. _____

What we have learned about the way we see things

See Student Book 5, pages 20–21

Home learning What I have learned...

Complete the crossword to help consolidate your learning from this module.

For advice on how to help the student with this activity, see page 7.

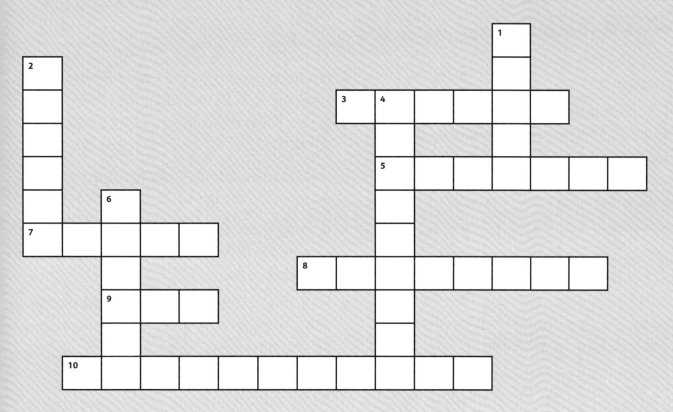

Across

3 This word describes materials that do not let light pass through them.
5 To bounce light off
7 The colour made when all the other colours of light are mixed
8 The ray travelling towards a mirror
9 The path of light shown as a line
10 This word describes materials that let light pass through them.

Down

1 The angle of incidence and the angle of reflection are this.
2 The dark shape made when light is blocked by an object
4 A device that uses mirrors to see around or over objects
6 A shiny object that lets us see objects reflected

2 Evaporation and Condensation

Extra support

Introduction

In this module students learn about the vital processes of condensation and evaporation and find examples of these in everyday life. Examples of evaporation include obtaining salt from seawater, and drying clothes. Examples of condensation include rainfall and dew.

The activities encourage students to think about evaporation and condensation on a huge scale. These processes, driven by energy from the Sun, form the water cycle. The water cycle is the way that water on Earth is recycled and not used up.

The next activities are designed to help students to understand that some substances dissolve in water and others do not. They learn that soluble materials dissolve in water to make solutions and that this process is called dissolving.

This module will help the student to practise these scientific enquiry skills:

- planning – asking questions and planning how to seek answers in a fair test (pages 22, 25, 27, 28, 29, 31, 32)
- predicting – stating what they think will happen and then comparing this with what actually happens (pages 22, 28)
- observation – collecting sufficient evidence, understanding the need for repeated observations or measurements (pages 22, 24, 25, 27, 28, 29, 30, 31, 32)
- recording – writing or drawing observations or stages in work and presenting results in line graphs or bar charts (pages 22, 24, 25, 27, 28, 29, 30, 31, 32)
- making comparisons – comparing sets of evidence or data (pages 22, 27, 28, 29, 32)
- drawing conclusions – examining results to identify any patterns and/or to suggest explanations (pages 22, 25, 27, 28, 32)
- evaluating results – evaluating the investigation and results and suggesting ways to improve the investigation (pages 22, 25, 26, 28).

Ways to help

Show the student examples of boiling, condensation, freezing and melting at home. You could also make solutions with the student. It is easy to dissolve salt or sugar in water. A common misconception is that substances disappear when they dissolve. Help the student to understand that the salt or sugar is still present in the water, by pouring some of the solution into a shallow dish and placing it in a warm place. When the water evaporates, the salt or sugar will remain in the dish, showing that it is still there in the solution.

Remember to encourage and help the student but try to let them find out as much as they can on their own.

Finally, help the students complete the 'What I have learned…' summary to test their understanding and recall.

Key words

condense/condensation, crystals, dissolve, evaporate/evaporation, filter, gas, insoluble, liquid, particles, solid, soluble, solution, states of matter, water cycle, water vapour

Scientific enquiry words

control, fair test, investigate, line graph, observe, plan, predict, record

Helping with activities

The following guidance is intended to offer advice to the parent, or other adult at home, on how to help the student with each home learning activity.

Find the missing properties (page 23)

Encourage the student to read out the content of the table and think about the missing information. Support them in drawing pictures of the particles in solids, liquids and gases in the boxes at the bottom of the page.

Examples of steam at home (page 24)

Remind the student that steam has a temperature much higher than that of boiling water. With the student, look around your home and local area for places where steam is used or produced. Help the student to research how steam is used in your country. Remind them about use of steam engines and locomotives in the past.

A model water cycle (page 26)

Provide a paper plate or a cardboard box and any modelling equipment available to help the student complete this activity. Encourage the student to think about the different parts of the water cycle. Help them to plan how they will model the water cycle and the materials they will use. They will take the model into school to be part of a display.

Condensation at home (page 27)

Start by asking the student to breathe on a clean mirror. Ask them what they can see on the mirror and where this has come from. Compare the condensation on the mirror that has been placed in the fridge and on the mirror that is at room temperature. What difference does the cold mirror make to the condensation produced?

Make crystals at home (page 29)

In order to produce good crystals the sugar or salt solution needs to be saturated (no more will dissolve), so you will need quite a lot. The water has to be warm to dissolve the solid. Make sure that the student is aware of the dangers of using warm water. Ensure that the student stirs the solution well before adding more of the sugar or salt. Ask them to explain what has happened to the water. Make sure the student keeps checking the dish of sugar or salt solution until all the water has evaporated. If you have a hand lens allow the student to observe the shapes of the crystals.

Using filters (page 30)

Talk to the student about the use of filters in general. Then discuss how filters are used in the home. Locate some different filters before the activity so that the student can find them easily.

Separation challenge (page 31)

Provide the student with a mixture of four or five different materials that have small grains and are safe to touch. If you have a magnet available, include some paper clips or other small magnetic objects in the mixture. Help the student to show the mixture to different people in your home.

What I have learned… (page 33)

This summary activity is designed to test the student's understanding and recall. Help the student to answer the questions, if needed.

Where does the water go?

See Student Book 5, pages 24–25

Class activity Investigate drying

This student is investigating how moving air affects **evaporation**. She blows moving air on one wet cloth with a hair dryer. The other wet cloth on the line is not in the moving air. This cloth is a **control**. The student will compare this cloth with the one in the moving air.

 How does the student test how much water each cloth has at the start of the investigation?

_____

 How does the student find out which cloth has lost the most water?

_____

Evaporation and drying

Plan and carry out an investigation to find out how you can get a cloth to dry out quickly. Check your plan with your teacher before starting your investigation.

Your teacher will tell you which condition to **investigate**.

- Wind – Does a wet cloth dry faster if it is in moving air?
- Temperature – Does a wet cloth dry faster in a warm place than in a cool place?
- Surface area – Does a wet cloth dry faster when spread out than when folded up?

 Predict what will happen. Write your prediction.

_____

Was your prediction correct?   Yes   No

 As a class, discuss your results and try to explain them using the idea of particles.

Where does the water go?

See Student Book 5, pages 26–27

Home learning Find the missing properties

The table shows the properties of the three **states of matter**. Fill in the gaps in the table.

For advice on how to help the student with this activity, see page 21.

Property	Solid		Gas
Volume		Fixed	
	Fixed		Spreads out to take the shape of the whole container
Density		Medium	
Ease of compression (squashing)	Very low		High
Ease of flow		Easy	

Draw the particle models for a **solid**, **liquid** and **gas**.

Solid

Liquid

Gas

Where does the water go?

See Student Book 5, pages 30–31

Home learning Examples of steam at home

You will have seen examples of steam being made at home and in your local area.

For advice on how to help the student with this activity, see page 21.

Steam survey

1. Survey your home and local area. Find five examples of steam being made. Do not heat water specially to make steam. **Observe** examples of steam that occur normally.

2. Complete the table.

Where I saw steam being made	How was the steam being made?

Research how steam is used in your country. Make a poster to take into school. Your poster will be part of a display on how we use steam.

Getting the water back

See Student Book 5, pages 32–33

Class activity Build your own water plant

A desalination plant is where water is **evaporated** from seawater and then **condensed** to give pure water.

 Make your own desalination plant.

1 Follow the instructions in the diagrams.

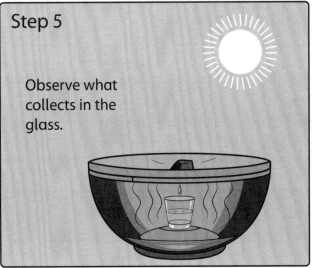

2 **Observe** what happens to the salt water in the bowl. **Observe** what collects in the glass.

Label the diagrams to show any places where evaporation is taking place.

How well did your desalination plant work? Discuss this with your teacher and classmates.

Do not taste the water you collected unless your teacher tells you it is safe.

Getting the water back

See Student Book 5, pages 34–35

Home learning A model water cycle

✋ Make a model of the **water cycle**.

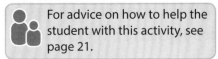
For advice on how to help the student with this activity, see page 21.

1. Think about which materials you will use.
2. Build your model.
3. Label the important parts of your model. Use the words in the word bank. You can add other labels too.

You could use: a cardboard box, a large piece of paper or card, or a paper plate for the background; and cotton wool or tissue paper, card, pipe cleaners or even paper mache for the parts of the water cycle.

| Evaporation | Clouds | Condensation | Rain |

💬 Use your model to explain the water cycle to people in your family.

Take your model to school. Your teacher will organise a big display.

Compare your model to other students' models.

💭 Have you missed anything from your model? How can you improve your model?

✏️ Write your ideas for improvements.

Dry and damp air

See Student Book 5, pages 38–39

Home learning Condensation at home

Investigate condensation.

You will need two small mirrors.

1. Ask an adult at home if you can place one of the mirrors in the fridge.
2. Place the other mirror on a table.
3. After ten minutes take the mirror out of the fridge. It should be very cold.
4. Place the cold mirror next to the other mirror on the table.
5. Breathe on both mirrors. Compare what happens.

For advice on how to help the student with this activity, see page 21.

What did you see on the cold mirror? _____

Explain why this happens. _____

_____

In the boxes, draw what the **particles** look like at both stages.

Boy's breath	Condensation

What is the process called when liquid water changes to **water vapour**?

_____

What is the process called when water vapour changes to liquid water?

_____

Which process do you see when you breathe on the cold mirror?

_____

Making solutions

See Student Book 5, pages 48–49

Class activity Investigate dissolving

Some solids **dissolve** in water. These solids are **soluble**.

Some solids do not dissolve in water. These solids are **insoluble**.

Investigate which solids are soluble and which are insoluble.

1 Answer the questions about how to make this a **fair test**.

Why must you use the same amount of solid for each test?

_____

Why must you use the same volume and temperature of water for each test?

_____

Why must you stir each mixture exactly the same number of times?

_____

2 **Predict** which solids you think will be soluble.

3 Try to dissolve each solid in water. Use clean water for each test.

4 **Observe** what happens. Does the solid dissolve or does it settle at the bottom of the beaker? Look at the mixture using a hand lens. Can you see any particles of the solid floating on the water or at the bottom of the beaker?

5 Complete the table.

Solid	Prediction	Was it soluble?	Was it insoluble?	What did the mixture look like?
Sand				
Sugar				
Talc				
Salt				
Chalk				

Which materials were soluble? _____

Were your predictions correct?   Yes   No

Making solutions

See Student Book 5, pages 50–51

Home learning Make crystals at home

 Make some **crystals**.

For advice on how to help the student with this activity, see page 21.

Show your family how to make crystals. Use sugar as your soluble solid.

1. Pour some warm water into a cup.
2. Slowly add sugar to the water and stir it well.
3. Keep adding sugar until no more will dissolve.
4. Pour your sugar **solution** into a shallow dish or plate and leave it in a warm sunny place.
5. **Observe** what happens to the solution.

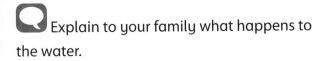 Explain to your family what happens to the water.

6. Look at any crystals that form. Use a hand lens if you have one.

 What shape are the sugar crystals? Draw the crystals in the box labelled 'Sugar'.

7. Repeat the investigation, but this time use salt.

 What shape are the salt crystals? Draw the crystals in the box labelled 'Salt'.

Sugar	Salt

8. Compare the sugar and salt crystals. How are they different?

_____

_____

Making solutions

See Student Book 5, pages 52–53

Home learning Using filters

Here are some examples of **filters**.

> For advice on how to help the student with this activity, see page 21.

Coffee filter

Tea bags

Vacuum cleaner filter

Dust mask

Sieve

Colander

Survey of filters

1 Survey your home to find as many different types of filter as you can.

2 **Record** your survey results in the table.

Example of a filter	Where I found it	What it is used to separate

Explain why a filter will separate rice from water but not salt from water. Use the words in the word bank in your answer.

_____

_____

| soluble | insoluble | dissolve | particle size | hole size |

Making solutions

See Student Book 5, pages 54–55

Home learning Separation challenge

1. Ask an adult to help you to mix together some different materials in a bowl, for example sand, salt, rice and paper clips.

 For advice on how to help the student with this activity, see page 21.

2. Challenge the people at home to separate the materials.

3. Ask them for ideas of how to do this. When they make a good scientific suggestion, let them try it out.

Use this checklist to help you.

	Yes	No
■ Did someone suggest using a sieve?	☐	☐
■ Did someone suggest using a magnet?	☐	☐
■ Did someone suggest adding the mixture to water?	☐	☐
■ Did someone suggest filtering?	☐	☐
■ Did someone suggest dissolving and then drying or evaporating?	☐	☐

4. Now demonstrate how a scientist would separate the materials.

✏️ Write your method. _____

Making solutions

See Student Book 5, pages 56–57

Class activity Testing saline solutions

 Testing saline solutions

You are going to **investigate** the amount of salt in three different saline solutions, labelled A, B and C. Your task is to find the correct solution for your local hospital.

> Remember: the saline solution in a saline drip used in hospitals should have 9 grams of salt in every 1000 cm^3.

1. Find how many grams an empty beaker weighs. **Record** the weight in your table.

2. Place the beaker on a warm, sunny windowsill and add 100 cm^3 of solution A. Label the beaker A.

3. Repeat this with two more beakers. Add 100 cm^3 of solution B to beaker B. Add 100 cm^3 of solution C to beaker C.

4. Wait until all the water has evaporated.

5. Find how many grams beakers A, B and C weigh now. This will be the beaker plus any salt it contains.

A saline drip is a solution of salt in water.

Beaker	Number of grams for the empty beaker	Number of grams for the beaker and the salt
A		
B		
C		

Calculations

The correct saline solution has 9 grams of salt in every 1000 cm^3.

How many grams of salt should 100 cm^3 of the solution contain? _____

To find the number of grams of salt, take the number of grams of the empty beaker away from the number of grams of the beaker plus the salt.

Grams of salt in 100 cm^3 of saline solution A = _____ .

Grams of salt in 100 cm^3 of saline solution B = _____ .

Grams of salt in 100 cm^3 of saline solution C = _____ .

The correct solution is saline solution _____ .

What we have learned about evaporation and condensation

See Student Book 5, pages 58–59

Home learning What I have learned...

Answer the questions to help review your understanding of this module.

Circle the correct answer to each question.

1. When puddles of liquid water dry up the process is called
   - a dissolving
   - b evaporating
   - c condensation
   - d filtering

2. Condensation is when
   - a steam is made
   - b liquid water becomes water vapour
   - c water vapour becomes liquid water
   - d a solid dissolves in water

3. In a solid the particles are
   - a far apart
   - b free to move around
   - c larger than other particles
   - d closely packed together

4. Gases have particles that are
   - a fixed and close together
   - b widely spaced and free to move
   - c smaller than the particles in solids
   - d invisible

5. The main source of energy that drives the water cycle is
   - a the Sun
   - b electricity
   - c clouds
   - d gravity

6. When a solid is heated it can form a liquid. This change of state is called
   - a evaporation
   - b boiling
   - c freezing
   - d melting

7. A solid that dissolves in a liquid is called
   - a insoluble
   - b a conductor
   - c soluble
   - d an insulator

8. To separate sand from water we can use
   - a dissolving
   - b filtering
   - c chromatography
   - d freezing

9. The gas formed from liquid water is
   - a carbon dioxide
   - b air
   - c water vapour
   - d condensation

10. The shape of liquids is always
    - a the shape of the lower part of the container
    - b the shape of the whole container
    - c never fixed
    - d rigid

3 The Life Cycle of a Flowering Plant

Extra support

Introduction

This module helps students to understand how plants reproduce. They will learn that plants have male and female parts and that flowers are pollinated by insects and other methods. Students will also learn that when pollen fertilises the ovum or egg, a seed forms. They will explore ways that seeds can be dispersed and why this is important.

Students will think about the life cycle of flowering plants and observe examples of plants at each stage in the cycle. They will learn how to identify and label the parts of a flower and identify the individual stages of a plant's growth.

This module will help students to practise these scientific enquiry skills:

- planning – asking questions and planning how to seek answers in a fair test (pages 38, 39, 41, 45)
- predicting – stating what they think will happen and then comparing this with what actually happens (page 45)
- observation – collecting sufficient evidence, understanding the need for repeated observations or measurements (pages 36, 38, 39, 40, 41, 44, 45)
- recording – writing or drawing observations or stages in work and presenting results in line graphs or bar charts (pages 36, 38, 39, 40, 41, 44)
- making comparisons – comparing sets of evidence or data (pages 36, 38, 39, 40, 45)
- drawing conclusions – examining results to identify any patterns and/or to suggest explanations (page 38, 40, 45)
- evaluating results – evaluating the investigation and results and suggesting ways to improve the investigation (pages 38, 41).

Ways to help

Point out examples of flowers and seeds. Supervise the student as they explore seeds found at home. If you have cut flowers in the house, or robust flowering plants in pots or a garden, allow the student to take one flower from each plant and observe it carefully. They could remove each part from the flower and stick it into a scrapbook or press the flowers and label the parts. Remind them never to pick wild flowers.

A useful way to learn about the life cycle of flowering plants is to plant seeds. This can be done with small pots placed on a windowsill. Encourage the student to observe carefully what happens, take notes over a period of time and water the plants.

Remember to encourage and help the student but try to let them find out as much as they can on their own.

Finally, help the students complete the 'What I have learned…' summary to test their understanding and recall.

> **Key words**
>
> | | | |
> |---|---|---|
> | anther | germination | reproduce/ |
> | carpel | insects | reproduction |
> | dispersal/ | life cycle | seed |
> | dispersing | male | seedling |
> | explosion | nectar | seed |
> | female | ovary | production |
> | (ovum) | plant | stamen |
> | fertilisation | pollen | stigma |
> | filament | pollen tube | style |
> | flowers | pollination/ | water |
> | fruit | pollinate | wind |

> **Scientific enquiry words**
>
> | | |
> |---|---|
> | evaluate | predict |
> | observe | test |

 ## Helping with activities

The following guidance is intended to offer advice to the parent, or other adult at home, on how to help the student with each home learning activity.

Label the life cycle (page 37)

Encourage the student to study the pictures, each of which shows one of the stages in the life cycle of a flowering plant. Talk to them about each stage and ask them what is happening.

Compare seeds (page 39)

Consider going on a mini-field trip so the student can see a wide variety of seeds. Encourage the student to realise that wind-spread seeds are very light and often have a 'parachute' or filamentous shape so they can catch the wind.

Flower survey (page 40)

Provide opportunities for the student to see a lot of different flowers. Point out that wind-blown pollen is likely to be on anthers that are held above the flower by long filaments. Insect-pollinated flowers encourage the insects to enter the flower to obtain nectar so the anthers are either deep inside the flower or overhanging where the insect is likely to land.

Design a flower (page 41)

Provide materials such as paper, coloured card, cotton wool and yellow powder paint. White powder such as talcum or flour would also work but may be harder to see. Encourage the student to think back to the flower survey so they can learn about flower design from nature.

Flower parts and what they do (page 43)

Encourage the student to read the functions of the flower parts carefully and then link the name of each flower part to the correct function. Help them to learn the terms by giving them fun tests and quizzes. If you have some cut flowers the student could make a dried flower display. Make sure they label all the flower parts.

Fertilisation (page 44)

The student will need plasticine or modelling clay to make a model carpel. If none is available, they can use card. Encourage the student to make three models to clearly show the different stages. The differences between the models will be in the position of the pollen grain and how far the pollen tube has grown down towards the ovary.

Plant life cycle display (page 46)

Encourage the student to make a colourful display of the plant life cycle. They will need paper plates and coloured pencils or pens. If paper plates are not available, the student can draw around a dinner plate to make card circles.

What I have learned… (page 47)

This summary activity tests the student's recognition and understanding of key terms from the module. Help the student to find the words in the word search, and then to write the function of the flower parts in their own words.

Flowering plants

See Student Book 5, pages 62–63

Class activity What are seeds like?

 Observe some **seeds**.

Your teacher will give you some samples of seeds.

1 **Observe** the outside of the seeds.

 Write three ways that the seeds are different.

1 _____   2 _____

3 _____

2 Your teacher will help you to cut the seeds. Look at the structure inside the seeds. Can you find two different types?

 Draw an example of each type in the boxes.

GOING FURTHER

Seeds of a plant can be one of two types; monocotyledonous (monocots) have a single cotyledon, and dicotyledonous (dicots) have two cotyledons.

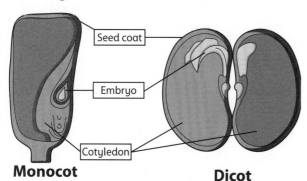

Monocot　　　**Dicot**

The seed coat protects the seed. The embryo will grow into a new **plant**. The cotyledon is the part of the embryo that will become the first leaves.

Did you find any monocots and dicots? _____

 Survey the plants near your school.

How many monocot and dicot plants can you find?

 How will you present your findings?

Other clues
- Monocot plants have **flowers** with petals in 3s, 6s or 9s. Dicot plants have flowers with petals in 4s or 5s.
- Monocot plants have leaves with veins in straight lines. The leaves of dicot plants have branched veins.
- Monocot plants have branched roots. Dicot plants have a long central root called a tap root.

Flowering plants

See Student Book 5, pages 64–65

Home learning Label the life cycle

Draw a line to join each label to the correct stage in the **life cycle** of a flowering plant.

 For advice on how to help the student with this activity, see page 35.

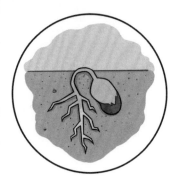   

| Young plant | Seedling | Adult plant | Seed |

Draw a large red arrow on the correct picture to show where **germination** takes place. Label this 'Germination'.

Draw the cycle in the correct order in the boxes. Label your drawings.

3 The Life Cycle of a Flowering Plant

Seeds, seeds everywhere!

See Student Book 5, pages 66–67

Class activity Make your own seed

✏️ Seeds need to spread out, away from the parent plant. Explain why.

_____

_____

The **wind** is an important way of spreading or **dispersing** seeds.

✋ Make a model seed.

Use one piece of paper. You can fold, cut and glue the paper, but you can only use one piece.

1 Design your seed so that it will fly as far as possible.

💭 Think about shapes that will blow easily in the wind. Look at some seeds that spread using the wind.

2 Make your seed.

3 Your teacher will help you to **test** your seed.

✏️ Draw your seed in the box above.

📏 Measure how far your seed flew.

My seed flew _____. Remember to include the units.

4 Compare your seed to your classmates' seeds.

✏️ Whose seed flew the longest distance? _____

How far did this seed fly? _____

Evaluate the seeds. Which features helped them to fly well?

_____

_____

Seeds, seeds everywhere!

See Student Book 5, pages 68–69

Home learning Compare seeds

Spread by wind

Spread by animals

Spread by explosion

Spread by water

 Compare different types of seeds.

For advice on how to help the student with this activity, see page 35.

1. Collect five samples of seeds at home or in your local area. Ask an adult to help you with this.

2. Complete the table. An example is done for you.

The plant the seeds come from	Spread by water	Spread by wind	Spread by explosion	Spread by animals
Dandelion		✓		

3. Compare the shapes of seeds spread by the wind with those spread by animals.

Draw an example of each type of seed. Label one difference between the seeds on each type.

Imagine lots of the same plants very close together.

 What are the plants competing for? _____

Why is it useful for the seeds to spread out away from the plant?

_____

_____

Pollinating flowers

See Student Book 5, pages 70–71

Home learning Flower survey

 How are flowers **pollinated**?

For advice on how to help the student with this activity, see page 35.

1 Look for different types of flowers in and near your home.

2 Try to decide which are pollinated by:

- wind
- **insects**
- other animals.

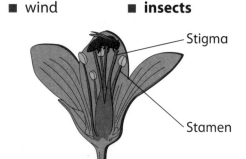

This flower attracts insects with scent, colour and sugary nectar.

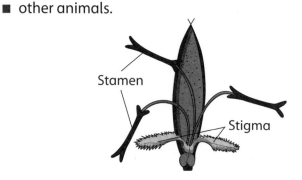

The pollen-holding stamens of this flower are high up so the wind can blow the pollen away.

- What clues do you get from the shape of the flowers?
- What clues do you get from the scent of the flowers?
- What clues do you get from the colours of the flowers?

Draw a flower you have seen insects visiting.

Label any parts of the flower that you think attract the insects.

Draw a flower you think is pollinated by the wind.

Label any features that you think help the wind to blow **pollen** away.

Pollinating flowers

See Student Book 5, pages 72–73

Home learning Design a flower

Wind-pollinated flower

Flower pollinated by a humming bird

 For advice on how to help the student with this activity, see page 35.

Flower pollinated by a bee

The bee orchid: one petal looks like a female bee

You will need: coloured paper or card; sticky tape; yellow powder paint, or white powder such as talcum or flour; kitchen sponges or cotton wool balls.

 How do some flowers attract insects and birds?

_____

Why do you think bees are attracted to the bee orchid?

_____

 Design and make your own flower.

1 Design your flower so that it is easy for a bee to pollinate it. Make your flower using paper or card.
2 The powder is the pollen. Put it into your flower.
3 Use the sponges or cotton wool balls to make your bee.
4 Place your bee on the flower to collect some **nectar**.

Can the bee collect nectar without getting powder (pollen) on it? If it can, you will need to redesign your flower or the plant will not survive.

 Draw your most successful flower design in the box.

Looking at flowers in detail

See Student Book 5, pages 74–75

Class activity The parts of a flower

✏️ Identify the parts of the flower in the pictures. Write the labels. You can use the words in the bank to help you.

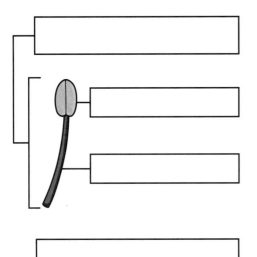

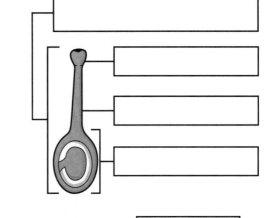

Stem
Petals
Style
Stigma
Carpel
Sepal
Ovary
Stamen
Anther
Filament

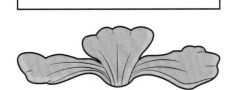

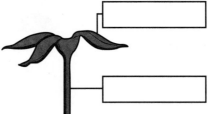

✏️ Draw a picture of the whole flower by drawing the different parts in the correct places.

Looking at flowers in detail

See Student Book 5, pages 74–75

Home learning Flower parts and what they do

Flowers are very important. They help the plant to **reproduce** by producing seeds.

 For advice on how to help the student with this activity, see page 35.

Draw a line to link each part of a flower to the job it does. One is done for you.

Petal	—	links the ovary and the stigma
Anther	—	protects the flower before it opens
Filament	—	contains the eggs and becomes the fruit with seeds
Ovary	—	the part that makes pollen
Sepal	—	the sticky part of the flower that receives pollen
Stigma	—	the male part of the flower
Style	—	holds the anther in place
Stamen	—	the bright part of a flower that attracts insects

(Petal is linked to "the bright part of a flower that attracts insects".)

Complete the sentences. You can use the words in the word bank to help you.

The **male** part of a flower is the **stamen**. It has two parts called the _____ and the _____.

The **female** part of a flower is the **carpel**. It has three parts called the _____, the _____ and the _____.

| ovary | anther | style | filament | stigma |

3 The Life Cycle of a Flowering Plant

Looking at flowers in detail

See Student Book 5, pages 76–77

Home learning Fertilisation

 Look at three different flowers at home or in your local area.

 For advice on how to help the student with this activity, see page 35.

 Draw the carpel of each flower and label the parts.

| Carpel of flower 1 | Carpel of flower 2 | Carpel of flower 3 |

 Draw the stamen of each flower and label the parts.

| Stamen of flower 1 | Stamen of flower 2 | Stamen of flower 3 |

Use modelling clay to make a model carpel and some pollen grains.

Slowly change your model to show people at home how the pollen grain grows a **pollen tube** and reaches the **ovum** or egg.

The big picture

See Student Book 5, pages 78–79

Class activity Do plants grow from seeds?

✋ Which part of a **fruit** grows into a new plant?

How can you prove that it is the seeds from a fruit that grow into new plants?

Work with a partner or in a small group.

1 Try planting different parts of a fruit and see whether they grow.

2 In four separate pots of soil or compost, plant:

  a pieces of the skin

  b pieces of the stalk

  c pieces of the core without seeds

  d the seeds.

✏️ **2 Predict** what you think the results of your investigation will be.

I **predict** that _____

_____ .

3 **Observe** and **water** the plant pots every day until you see some growth.

✏️ Do your results prove the role of seeds in making new plants?

_____

_____

The big picture

See Student Book 5, pages 80–81

Home learning Plant life cycle display

We can use paper plates to make some very interesting displays.

✏️ **1** Write each of the stages in a plant's life cycle on a separate paper plate. Use the words in the word bank.

Pollination	**Seed production**
Seed dispersal	**Fertilisation**
Flower production	**Seedling**
Germination	

👥 For advice on how to help the student with this activity, see page 35.

If you do not have paper plates, you can draw around a dinner plate on card or paper. Then cut out the circles.

2 Draw a picture of each stage on the plate.

3 Arrange the plates on a table or work surface so they show the life cycle of a flowering plant.

4 Cut some arrows out of paper or card. Use the arrows to link the plates in a circle.

5 You could fix the whole display to a large piece of paper.

💬 Use your display to explain the cycle to people in your family.

Take your display to school. Your teacher will make a class display.

✏️ Imagine you are a seed. Write a short story about your life until you are a fully grown plant and making your own seeds.

_____
_____
_____
_____
_____
_____

What we have learned about the life cycle of a flowering plant

See Student Book 5, pages 82–83

Home activity What I have learned...

In the word bank there are words from this module. Find them in the wordsearch. Words go across, down and diagonally.

 For advice on how to help the student with this activity, see page 35.

anther	stigma	style
fruit	ovary	germination
pollination	sepals	fertilisation
flower	seeds	
petals	filament	

```
g e r m i n a t i o n c o a i
t f o s j s e p a l s v m n n
d e p c t e v y b z r l r t n
d r o t g i w j f y h e n h b
l t w f p m g s a v s o i e z
i i b l c f l m k n i e j r d
g l d o c a j z a t m b e k w
g i f w t s e g a j f t z d m
j s o e u t n n z k v f l i s
f a p r d y i p o v a r y b c
t t z t k l w h z r j u y a w
e i i v l e a y e a o i l j o
s o w o x s b o u j n t z h j
n n p g l o u f i l a m e n t
```

✏️ Describe what each of the following parts of a flower does.

Anther _____

Stigma _____

Ovary _____

4 Investigating Plant Growth

Extra support

Introduction

The main purpose of this module is to help students to learn about the conditions needed for healthy plant growth.

This module provides opportunities for students to analyse data and interpret the results of investigations. Many of the tasks involve observation and recording over a period of days or even weeks as the student has to wait for the plants to grow, so encourage them to be patient.

This module will help students to practise these scientific enquiry skills:

- planning – asking questions and planning how to seek answers in a fair test (pages 50, 56, 57)
- predicting – stating what they think will happen and then comparing this with what actually happens (pages 50, 57)
- observation – collecting sufficient evidence, understanding the need for repeated observations or measurements (pages 50, 51, 54, 57, 58)
- recording – writing or drawing observations or stages in work and presenting results in line graphs or bar charts (pages 50, 51, 52, 53, 54, 57, 58, 59, 60)
- making comparisons – comparing sets of evidence or data (pages 50, 51, 53, 54, 57, 58, 59, 60)
- drawing conclusions – examining results to identify any patterns and/or to suggest explanations (pages 50, 53, 57, 58, 59, 60)
- evaluating results – evaluating the investigation and results and suggesting ways to improve the investigation (pages 52, 57, 58).

Ways to help

Encourage the student to observe plants on their way to school or when they go on visits with you. Ask them to look at plants growing in shady areas and especially to notice areas where plants are growing well and areas where there are very few plants. Point out barren areas of ground and ask why no plants are growing. Elicit that there is little water or not much light or barren soil or sand with few nutrients for the plants.

Provide opportunities for the student to grow plants at home and observe which conditions help the plants to thrive.

Remember to encourage and help the student but try to let them find out as much as they can on their own.

Finally, help the students complete the 'What I have learned…' summary to test their understanding and recall.

Key words

growth seeds
light/sunlight warmth
plant water

Scientific enquiry words

conclusion/ factor predict
  conclude fair test present
control line graph   results
  (variable) measure record
evaluate observe variable

Helping with activities

The following guidance is intended to offer advice to the parent, or other adult at home, on how to help the student with each home learning activity.

Survey of places where plants grow (page 51)

Take the opportunity to discuss what conditions plants need in order to grow. Take the student outside to find unusual places where plants manage to survive.

Presenting data (page 53)

Recording results and presenting data are important scientific enquiry skills. Find examples of graphs and charts used in everyday life, such as in newspapers, magazines, the Internet and on television. Demonstrate how presenting data in graphs and charts makes interpretation much easier. Students often find drawing the axis and selecting an appropriate scale difficult.

How plants use their different parts (page 55)

Look at any plants growing in and around the home. Talk about the importance of each part of the plant. Encourage the student to think about what the plant uses each part for, and to deduce that the position of a plant part may be a clue to its use. For example, a flower would not work well under soil because it attracts flying insects.

What are variables? (page 56)

Identifying variables is a key scientific enquiry skill. Encourage the student to think about controlling variables. For example, if they are measuring the height of a stem, they should make sure that they always place the zero on the scale at the bottom of the stem. If they are measuring the temperature of the place where a plant is growing, they should control whether a door or window is left open.

Plants need light to grow (page 58)

Discuss the student's school investigation on plants and sunlight. Discuss whether the students could use the same investigation set-up to test whether plants need water. The answer is that they would need to use a different set-up because the variables would be different.

Rules for drawing graphs and charts (page 59)

This activity and the next activity are a culmination of all the learning about presenting results or data. Encourage the student to add to the list of rules. There may be a rule that is personal to them. They could rewrite the rules and display them to help them to remember.

Drawing graphs and charts (page 60)

Look out for one of the key errors when drawing graphs, which is the incorrect use of scales. Some students merely list the numbers from the data rather than constructing a proper scale.

What I have learned… (page 61)

This summary activity allows the student to show their understanding of the conditions that help plants to grow well, as well as to apply their knowledge about presenting data and results as graphs and charts. Help the student to find any answers that they do not know.

Investigating seed germination

See Student Book 5, pages 86–87

Class activity What do seeds need to grow well?

✋ What conditions do **seeds** need to grow well?

Your teacher will give you some seeds.

1 Split your seeds into two equal sets.

2 Your teacher will take you outside. Look around the area for places to **plant** the seeds.

💭 Think about what seeds need in order to grow and be healthy.

3 **Predict** where you think the seeds will grow best. **Predict** where you think the seeds will not grow at all. Explain your predictions.

✏️ I **predict** the seeds will grow best _____

because _____ .

✏️ I **predict** the seeds will not grow _____

because _____ .

4 Plant one set of seeds in the place where you **predict** they will grow best.

5 Plant one set of seeds in a place where you **predict** they will not grow at all.

6 **Observe** your seeds for three weeks.

7 Draw a table to **record** your predictions *and* your observations.

✏️ Were your predictions correct?   Yes   No

💭 Compare your findings with your classmates' findings.

Investigating seed germination

See Student Book 5, pages 86–87

Home learning Survey of places where plants grow

 Talk to the people at home about how **plants** grow.

Tell them why plants need **sunlight**.

What other things do plants need in order to grow well?

 Survey of plants growing in unusual places

Carry out a survey of plants growing in unusual places in your local area.

> Ask an adult to go with you. Do not explore on your own.

1 Find plants growing in unusual or unexpected places.

 How are the plants surviving? Talk about how they get all the things they need for them to live.

2 Draw pictures or take photographs of the plants.

Which is the most unusual place where you see a plant growing? Draw a picture or stick a photograph in the box.

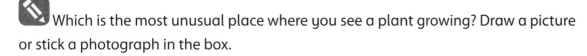

3 Take your drawings or photographs into school.

4 Compare them with your classmates' drawings and photographs.

 Decide who found the plant that was growing in the most unusual place.

> For advice on how to help the student with this activity, see page 49.

Investigating seed germination

See Student Book 5, pages 88–89

Class activity Investigation support

How important are **water** and **warmth** to seed germination? Some students carried out this investigation. They planted grass seeds in four pots – ten seeds in each pot. They put the same amount of soil in each pot. They placed the pots in different conditions and **recorded** the results in a table.

	Seedling height (cm)			
	Pot 1 water and warmth	Pot 2 water, no warmth	Pot 3 no water, warmth	Pot 4 no water, no warmth
Week 1	1	0	1	0
Week 2	3	1	1	0
Week 3	5	1	0 (Drooped)	0
Week 4	6	2	0 (Dead)	0

Choose the type of graph that will **present** the students' **results** in the best way. Draw the graph on the graph paper.

Evaluate the investigation:

1 Was it a **fair test**? Explain your answer. _____

_____

2 What can you change to improve the investigation? _____

_____

Investigating seed germination

See Student Book 5, pages 88–89

Home learning Presenting data

Talk to the people at home about how we can **present** data and **results**. Look for examples of data that is presented in newspapers, on television and on the Internet.

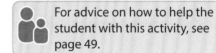

For advice on how to help the student with this activity, see page 49.

- Discuss why the data is presented in this way.
- Does it make the results easier to understand?
- Does it make it easier to read the results?

This table gives the results of another group of students for the investigation 'How important are water and warmth to seed germination?' The conditions for pots 1–4 are the same as on page 52.

	Seedling height (cm)			
	Pot 1	Pot 2	Pot 3	Pot 4
Week 1	2	1	1	0
Week 2	4	2	1	0
Week 3	5	2	0 (Dead)	0
Week 4	7	3	0 (Dead)	0

Is it easy to see from this table whether there are any trends or patterns in the results?

Draw a suitable graph to **present** the **results**.

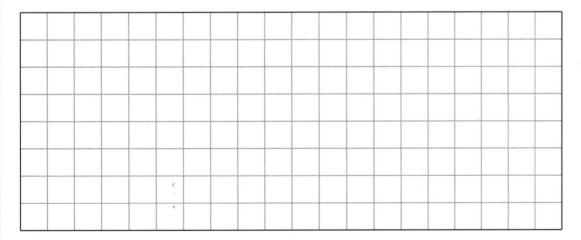

Is it easier to see trends or patterns using the graph or the table? _____

What can you **conclude** from these results? _____

Let there be light!

See Student Book 5, pages 90–91

Class activity Fieldwork

Measure and **observe** plants.

Your teacher will take you outside to carry out some fieldwork.

1 Find three different plants.

2 Use a ruler to **measure**:
- the length of the stem
- the width of the stem
- the length of a leaf at the top of the plant
- the length of a leaf at the bottom of the plant.

3 Count the number of petals on one of the flowers and **observe** their colour.

Record your measurements and observations in the table. Remember to include the units.

	Length of stem	Width of stem	Length of leaf at the top of the plant	Length of leaf at the bottom of the plant	Number of petals	Colour of petals
Plant 1						
Plant 2						
Plant 3						

4 Your teacher will show you the roots of some plants.

Draw the roots.

Describe how the plants are the same. _____

_____

Describe how the plants are different. _____

_____

Let there be light!

See Student Book 5, pages 90–91

Home learning How plants use their different parts

Show the diagram of the plant to someone at home. Talk about the different parts of the plant. What are the different parts used for?

For advice on how to help the student with this activity, see page 49.

Complete the table by writing the uses of each of the parts of a plant.

Part of plant	Uses
Flower	
Leaf	
Stem	
Root	

Use the table to help you complete the following sentences.

Plant roots are important because they _____.

Leaves are important because they _____.

Stems are important because they _____.

Flowers are important because they _____.

Let there be light!

See Student Book 5, pages 92–93

Home learning What are variables?

Talk to the people at home about how we carry out investigations.

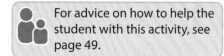
For advice on how to help the student with this activity, see page 49.

Discuss the **variables**. What is a **control variable**?

Here are examples of three investigations that students have carried out.

Investigation 1: Students planted ten grass seeds in four pots with equal amounts of soil. They placed the pots in different conditions:

- Pot 1: water and warmth
- Pot 2: water and no warmth
- Pot 3: no water and warmth
- Pot 4: no water and no warmth.

They **measured** the height of the seeds every week on the same day.

Investigation 2: Students planted five cress seeds in four pots. They placed the pots in different conditions:

- Pot 1: water and soil
- Pot 2: water and no soil
- Pot 3: no water and soil
- Pot 4: no water and no soil.

They counted how many seeds were growing every day.

Investigation 3: Students put different numbers of sunflower seeds in four pots of soil. They placed the pots in the same conditions:

- Pot 1: one seed – water and warmth
- Pot 2: two seeds – water and warmth
- Pot 3: five seeds – water and warmth
- Pot 4: ten seeds – water and warmth.

They **observed** the seed **growth** every week.

Use different colours to circle these **factors** for each investigation.

- Circle the **control variable** (what is being kept the same) in red.
- Circle the **variables** (what is varying) in green.
- Circle what is being **measured** in blue.

Let there be light!

See Student Book 5, pages 94–95

Class activity Investigate the effect of light on plant growth

Do plants grow better in more sunlight?

Your teacher will give you four boxes.

1 Cut out the front of each box to make a big hole. Label the boxes 1, 2, 3 and 4.

| **Box 1:** Leave the hole open. | **Box 2:** Cover the hole with one layer of tissue paper. | **Box 3:** Cover the hole with five layers of tissue paper. | **Box 4:** Cover the hole with thick black paper. |

2 Put the same number of young seedlings inside each box. They should all be the same type of plant. They should all be about the same size and look healthy.

3 Place all four boxes on the same windowsill with the holes facing towards the sun.

4 Give the same amount of water to each plant every day.

5 **Predict** which plants will grow the best.

6 At the end of the investigation, **measure** and **observe** the plants. **Record** your results in the table.

- **Measure** and **record** the height of all the plants in each box.
- Work out the average height of the plants in each box.
- **Observe** what the plants look like, including their colour.

> To work out the average, add up all the results. Divide the total by the number of results you have.

Box	Height of all the plants (cm)	Average height of the plants (cm)	Observations	Colour of the plants
1				
2				
3				
4				

Was your prediction correct?   Yes   No

What is your **conclusion** about the effect of **light** on seed growth?

_____

Why did the seedlings have to be the same plant type and the same size?

_____

Let there be light!

See Student Book 5, pages 94–95

Home learning Plants need light to grow

💬 Talk to the people at home about the investigation you did at school. Explain how you know that plants need light as well as water and warmth to grow.

 For advice on how to help the student with this activity, see page 49.

✋ Do plants grow better in more sunlight?

Some students have carried out this investigation. They placed seedlings in four different boxes. The pictures show the plants after two weeks.

A **B** **C** **D**

✏️ Can you match up each plant with the box that it was in? Write a reason for each answer.

Box 1: The hole is uncovered.

I think plant _____ was grown in Box 1.

My reason is _____
_____.

Box 2: The hole is covered with one piece of tissue paper.

I think plant _____ was grown in Box 2.

My reason is _____
_____.

Box 3: The hole is covered with five pieces of tissue paper.

I think plant _____ was grown in Box 3.

My reason is _____
_____.

Box 4: The hole is covered with a piece of thick black paper.

I think plant _____ was grown in Box 4.

My reason is _____
_____.

💬 Can you use the same investigation set-up to test whether plants need water?

Scientific enquiry: Graph and chart practice

Home learning Rules for drawing graphs and charts

Why it is important to **present** your **results** in graphs and charts?

For advice on how to help the student with this activity, see page 49.

Some rules for drawing graphs and charts

Charts and graphs can help us to see data more clearly. Not every type of chart shows everything we need. You have to decide which type is best for your set of results.

1 First decide which would be the best graph or chart to use.

- If only one of the two sets of data is numbers use a bar chart, for example chart A.
- If both the sets of data are numbers use a **line graph**, for example graph B.

A
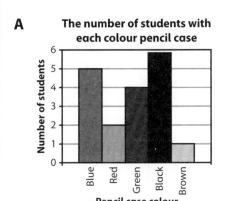

B
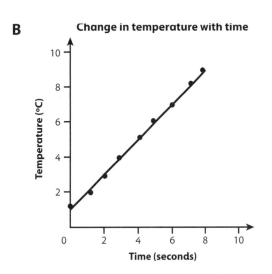

2 Choose a scale that is big enough to plot all the results. A big scale makes it much easier to plot the graph accurately. It also makes the graph easier to read.

3 Label the axes (x and y). The x-axis is the horizontal line. The y-axis is the vertical line.

4 Make sure your numerical scale increases by equal amounts, for example 5 or 10 for each square on the graph paper.

Write your own rule for drawing graphs and charts.

5 _____

_____

Scientific enquiry: Graph and chart practice

Home learning Drawing graphs and charts

Decide on the best way to **present** the **results** in tables A and B. Draw your graphs or charts on the graph paper. Use the rules on page 59 to help you.

A

Birth month	Jan	Feb	Mar	Apr	May	Jun	Jul	Aug	Sep	Oct	Nov	Dec
Number of students	3	5	1	0	7	9	4	11	9	10	2	1

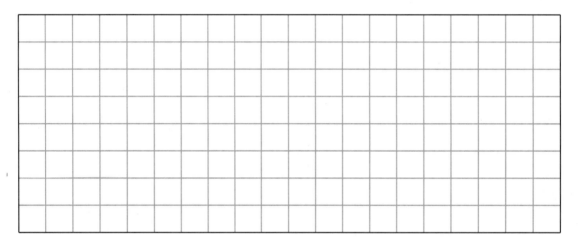

B

Age of student	2	4	6	8	10	12
Height of student (cm)	81	96	108	120	123	127

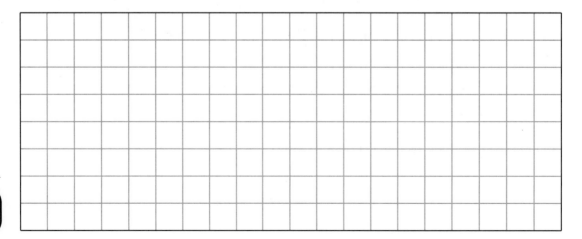

What we have learned about investigating plant growth

See Student Book 5, pages 98–99

Home activity What I have learned...

Complete the following tasks to review your learning from this module.

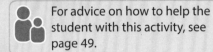
For advice on how to help the student with this activity, see page 49.

1. Which of these conditions will help a plant to grow the best? Circle the correct answer.

   Warmth and water   Warmth and no water   No warmth or water

2. Which of these conditions will help a plant to grow the best? Circle the correct answer.

   No sunlight and water   Sunlight and water   No sunlight and no water

3. These results show the number of flowers on one example of the same type of plant. Calculate the average number of flowers for this type of plant. Show your working out.

Plant	A	B	C	D	E	F	G
Number of flowers	3	1	1	0	2	3	4

4. Write two reasons why we display results in graphs and charts.

   1. _____
   2. _____

5. What is the best way to **present** the following **results**? Explain your answers.

   a. | Flower colour | Number of flowers |
      |---|---|

   The best way to **present** these **results** is _____
   because _____.

   b. | Plant height | Number of plants |
      |---|---|

   The best way to **present** these **results** is _____
   because _____.

5 Earth's Movements

Extra support

Introduction

The main purpose of this module is to help students to understand that the Earth orbits the Sun and the Moon orbits the Earth. A common misconception is when students infer from their daily observations of the *apparent* movement of the Sun that it is the Sun that moves.

Students will learn that the Earth completes a full spin on its axis every 24 hours and that this gives us night and day. They will also learn that it takes 365¼ days for the Earth to orbit the Sun. This gives us a year. Finally, students will investigate and learn about the lives and discoveries of some scientists who have studied the solar system. They will carry out their own research to find out about the planets of the solar system.

This module will help students to practise these scientific enquiry skills:

- planning – asking questions and planning how to seek answers in a fair test (pages 66, 68, 71)
- predicting – stating what they think will happen and then comparing this with what actually happens (pages 68, 71)
- observation – collecting sufficient evidence, understanding the need for repeated observations or measurements (pages 64, 65, 66, 67, 68, 70, 71)
- recording – writing or drawing observations or stages in work and presenting results in line graphs or bar charts (pages 66, 67, 68, 69, 70, 71, 72)
- making comparisons – comparing sets of evidence or data (pages 67, 68, 69, 70, 71, 72)
- drawing conclusions – examining results to identify any patterns and/or to suggest explanations (pages 66, 67, 68, 71)
- evaluating results – evaluating the investigation and results and suggesting ways to improve the investigation (page 71).

Ways to help

To help the student with the misconception that the Sun moves and the Earth stays still, you can show how an object can seem to move but is in fact not moving. An effective way to do this is to slowly turn the student around in the middle of a room. A table will seem to be moving as they slowly spin. You can also point out during a bus or car journey that, for example, trees, houses and buildings seem to be rushing by, but they cannot be moving.

Use a torch and a ball to model the Earth's spin. This will help the student to understand day and night.

Remember to encourage and help the student but try to let them find out as much as they can on their own.

Finally, help the students complete the 'What I have learned…' summary to test their understanding and recall.

Key words

24 hours	Moon	spin
axis	night	stars
day	orbit	Sun
Earth	planets	year

Scientific enquiry words

conclude/ conclusion	evidence	present
	observe	research
evaluate	predict	share

Helping with activities

The following guidance is intended to offer advice to the parent, or other adult at home, on how to help the student with each home learning activity.

Modelling the Earth's movements (page 65)

This activity is designed to help the student understand the complex way in which the Earth moves. It also reinforces the learning that the Sun does not move across the sky but the Earth moves around the Sun.

The Sun appears to move across the sky (page 67)

This activity requires a large space to move around in. Provide the student with a large piece of paper or card and a marker pen. The activity will help the student to realise that sometimes objects appear to move when it is actually the person that is moving. This activity should help them to understand how the Sun appears to move across the sky.

Measuring shadows (page 68)

This activity will require the student to be outside in sunlight. Ensure that they are properly protected from the Sun. This activity reinforces the understanding that the changing position of the Earth in relation to the Sun means that shadows change in length and direction throughout the day. The student measures the shadow cast by the same object or person at different times during the day to prove this. Make sure the object or person is in exactly the same position each time.

Investigating sunlight hours (page 69)

This activity is intended to help the student to realise that the amount of sunlight we receive varies during the Earth's yearly orbit of the Sun, depending on the tilt of the Earth's axis in relation to the Sun.

Keep a morning and night diary (page 70)

Encourage the student to keep a record of the different activities that they do in the morning and at night for one week. Also ask them to sit quietly and reflect on the sights, sounds and smells that they observe in the morning and again at night when it is dark.

Plotting graphs (page 72)

This is an opportunity for the student to practise plotting data. A bar chart would work best for this data. Talk about the fact that choosing to plot the data in Earth days will mean that the numbers have a large range, which will make the chart difficult to plot. Help the student to plot the chart using Earth years.

A postcard from another planet (page 74)

This is an opportunity to encourage literacy skills and for the student to summarise their understanding of the solar system and the conditions on the planets in a fun way. Encourage the student to use their imagination, but also to base their writing on the facts they have found about one of the planets.

What I have learned... (page 75)

This summary activity tests the student's recognition and understanding of key terms from the module. As the student finds each word, ask them to tell you what the word means and what they have learned about it.

The Sun appears to move, but it doesn't

See Student Book 5, pages 102–103

Class activity Model the movement of the Earth and Moon

Work in a group of between three and five people.
Your teacher will take you into a large open space.

You could make different masks, hats or cards to show what role you are playing. For example, if you are the Sun you could have a large star.

Model the movements of the **Earth** and **Moon**. Your group is going to model:

- how the Earth rotates
- how the Earth **orbits** around the **Sun**
- how the Moon orbits around the Earth.

1. One person will model the Sun.
   - Use a torch to model the light energy of the Sun. Stand still in the centre of the space.

2. One person will model the Earth.
   - Carefully **spin** on your **axis** (on the spot). It takes the real Earth **24 hours** (one **day**) to do this.
   - Orbit around the Sun while spinning on your own axis. It takes 365¼ days to do this. This means you will move more slowly around the Sun than you spin on your axis.

3. One person will model the Moon.
   - Orbit around the Earth. It takes the real Moon 28 days to do this. Try to move around the Earth at the right speed in relation to the movement of the Earth.
   - Hold a mirror to reflect the light from the Sun.

Discuss these questions in your group or with the whole class.

- How did it feel to be the Earth? Did you feel dizzy with all the spinning and rotating?
- How did it feel to be the Moon moving around the Earth?
- Was it easy to move at the correct speed in relation to the other objects? Did you bump into each other?

The Sun appears to move, but it doesn't

See Student Book 5, pages 102–103

Home learning Modelling the Earth's movements

Look at the diagram of the Earth orbiting the Sun.

For advice on how to help the student with this activity, see page 63.

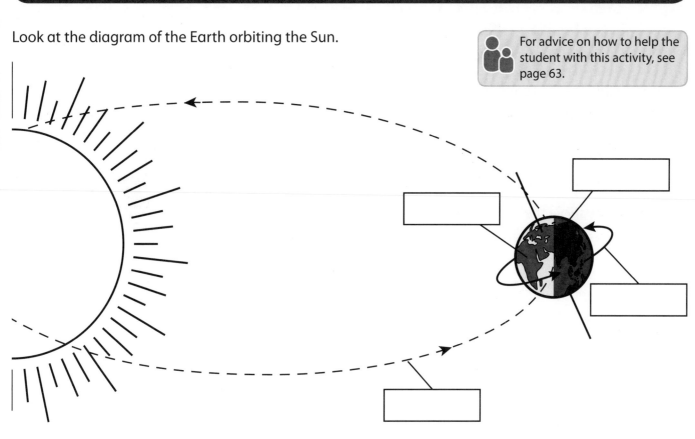

Talk to the people at home about the diagram.

Using the words in the word bank, fill in the empty label boxes.

| Daytime | Night-time | One day | One year |

GOING FURTHER

Draw the Moon in any position close to the Earth. Colour the Moon to show which parts are in sunlight and which parts are in shade.

5 Earth's Movements

The Sun appears to move, but it doesn't

See Student Book 5, pages 104–105

Class activity Moving Sun?

 How does the Sun appear to move across the sky?

Look at the picture. You are going to track the Sun as it appears to move across the sky during the day.

1. Start the investigation early in the morning.
2. Look at the classroom clock. Write the time on a sticky note.
3. Carefully **observe** the position of the Sun through the window. Place your sticky note on the window to mark the position of the Sun.
4. Repeat this every 30 minutes throughout the school day. Make sure you stand in exactly the same place each time so that your results are reliable.
5. At the end of the day, **observe** the positions of the sticky notes on the window.

> You will need some sticky notes.

> Remember: do not look directly at the Sun or the very bright light could damage your eyes.

Draw the shape that the sticky notes have made.

What shape have the sticky notes made? _____

 What is your **conclusion**? Has the Sun appeared to move across the sky?

Has the Sun really moved across the sky? _____

> Remember: our **planet** is rotating and moving around the Sun. The Sun is still. It doesn't move across the sky – it just appears to do so.

The Sun appears to move, but it doesn't

See Student Book 5, pages 104–105

Home learning The Sun appears to move across the sky

Explain to the people at home why some people think that the Sun moves across the sky. Tell them that you have found out that the Sun doesn't actually move.

For advice on how to help the student with this activity, see page 63.

Demonstrate how the Sun appears to move.

In this activity you will model the movement of the Earth and the object you track will represent the Sun.

Make sure you have lots of space. Move anything that might get broken.

1 Stand in the centre of the room.
2 Hold up a large piece of paper in front of you.
3 Choose an object in the room such as a chair.
4 Turn on the spot until the object is lined up with the top left corner of the paper.
5 Draw a dot on the paper to show the position of the object. Look at diagram A.
6 Make a small turn anticlockwise on the spot. Draw another dot on the paper to show the position of the object. Look at diagram B.
7 Keep turning in the same direction. After each small turn, draw another dot on the paper.
8 Continue until the object is lined up with the top right corner of the paper.

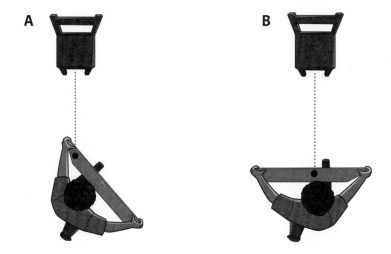

What moved in this investigation? Was it you or the object? _____

Remember that you represent the Earth and the object represents the Sun.

The Sun appears to move, but it doesn't

See Student Book 5, pages 106–107

Home learning Measuring shadows

💬 Talk to the people at home about what a shadow is.

- How are shadows formed?
- What happens to the light when it hits an object?
- Do all objects make good shadows?

✋ How do shadows change throughout the day?

You are going to measure shadows made by the Sun at different times of day. Choose whether to measure the shadow made by a building, a person or another large object.

👥 For advice on how to help the student with this activity, see page 63.

Carry out this investigation outside on a sunny day.

Make sure everyone is wearing sunscreen or has their skin covered. Why should you wear sunscreen?

📏 **1** Measure the length of the shadow in the morning, before you go to school.

✏️ Draw what the shadow looks like in the box on the left. Write the time. Write the length of the shadow.

📏 **2** Measure the length of the shadow when you get home from school.

- Make sure the object or person creating the shadow is in the same place as before.

✏️ Draw what the shadow looks like in the box on the right. Write the time. Write the length of the shadow.

The shadow before school   Time: _____

Length: _____

The shadow after school   Time: _____

Length: _____

4 Look at the two drawings and the measurements. Compare the shadows.

✏️ Explain why the shadows are different. _____

_____

The Sun appears to move, but it doesn't

See Student Book 5, pages 108–109

Home learning Investigating sunlight hours

Talk to the people at home about the sunrise and sunset where you live.

- What time does the Sun rise?
- What time does the Sun set?
- Do these times change during the year?

For advice on how to help the student with this activity, see page 63.

Ask someone at home to help you with this research project.

Find out the sunrise and sunset times for your local area. You could use local newspapers or the Internet to do this.

Record your findings in the table.

Month	Sunrise time	Sunset time	Total hours of sunlight
January			
February			
March			
April			
May			
June			
July			
August			
September			
October			
November			
December			

Use your table to answer the questions.

Which month has the longest days? _____

Which month has the shortest days? _____

5 Earth's Movements

How long does it take the Earth to spin on its axis?

See Student Book 5, pages 110–111

Home learning Keep a morning and night diary

Talk to the people at home about day and night.

For advice on how to help the student with this activity, see page 63.

- Ask them how they know it is day. What can they hear and see?
- Ask them how they know it is night. What can they hear and see?

Keep a diary for a week to record what you do in the morning and at night.

Day of the week	Things you do in the morning	Things you do at night
Sunday		
Monday		
Tuesday		
Wednesday		
Thursday		
Friday		
Saturday		

Compare the things you do in the morning to the things you do at night.

What is the same? _____

What is different? _____

When it is dark, how are the sounds around you different? _____

_____

Compare the things you do at night and in the day to the things other people at home do.

Are there any differences?

How long does it take the Earth to spin on its axis?

See Student Book 5, pages 110–111

Class activity Investigating the Earth's tilt

Is the Earth always half in sunlight and half in darkness?

Work with a partner to model the Sun shining on the Earth.

> You will need: a torch and a tennis ball with a skewer or piece of dowel pushed through it.

- Use the torch to represent the Sun.
- Use the tennis ball to represent the Earth.
- The skewer or piece of dowel represents the axis through the Earth.

1 Make some predictions.

Predict whether the parts of the Earth that are in light and in darkness will be equal or not.

Predict what will happen if you change the tilt of the Earth. Can you increase the area of the Earth that is in sunlight by altering the Earth's axis?

2 Shine the torch at the ball. How much of the ball is lit up?

How can you measure the area that is lit?

- You could draw around the ball on graph paper.
- On the graph paper, colour the area of the ball that is lit.
- Then count the coloured squares.

This will give you an estimate for the area of the ball that is lit.

3 Can you alter the area of the Earth that is in sunlight by changing the tilt of the axis?

- Hold the axis and hold the ball in front of the Sun.
- Tilt the ball backwards then forwards. Does this change the area of the ball that is lit up?

4 Look at your results. What do you **conclude** from this investigation?

Were you able to increase the amount of light that reached the ball by changing its tilt?

How does the Earth orbit the Sun?

See Student Book 5, pages 112–113

Home learning Plotting graphs

You are going to practise plotting data on a graph or chart.

💬 Talk to the people at home about graphs and charts.

- Why do we use graphs and charts?
- Why don't we just use tables to **present** our results and data?

For advice on how to help the student with this activity, see page 63.

✏️ Choose the best way to **present** the data in the table. It shows the length of the **year** for each planet of our solar system. Plot your graph or chart on the graph paper.

Planet	Length of year in Earth days (and Earth years)
Mercury	88 (0.2 Earth years)
Venus	225 (0.6 Earth years)
Earth	365 (1 Earth year)
Mars	687 (2 Earth years)
Jupiter	4331 (12 Earth years)
Saturn	10 747 (29 Earth years)
Uranus	30 589 (84 Earth years)
Neptune	60 225 (165 Earth years)

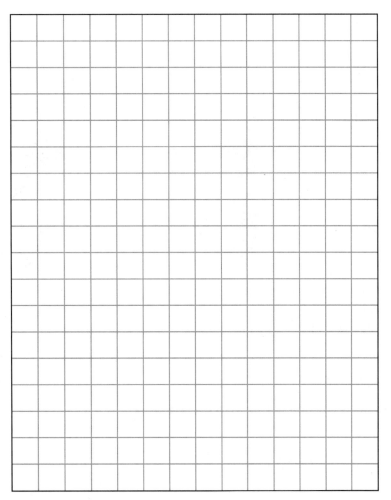

✏️ How old would you be in Jupiter years? _____

How old would you be in Mercury years? _____

The discovery of our solar system

See Student Book 5, pages 116–117

Class activity Planet research project

You have completed lots of work on the solar system and the **planets** it contains.

A model of the solar system

 Research one planet in our solar system.

1. Choose a planet in our solar system for your research project.
2. **Research** your planet. Use the Internet, books, your work from this module and any information your teacher gives you.
3. Use a paper plate to represent your planet.

Colour in your plate so that it looks like your planet.

4. From your research find some facts about your planet that you want to **share**.

Write the facts on the back of your plate. Some important facts to include are the answers to these questions.

- What is your planet's name?
- What is the surface temperature?
- What is this planet's distance from the Sun?
- Is the planet sunny or dark?
- What is the length of a year for this planet?
- What size is the planet?
- What is the mass of the planet?
- Does it have atmosphere or not?

Add any other interesting information you have found out.

5. Display your planet model in the classroom.
   - Your teacher may hang up the planets to create a model of the solar system.

The discovery of our solar system

See Student Book 5, pages 116–117

Home learning A postcard from another planet

💬 Talk to the people at home about the solar system.

For advice on how to help the student with this activity, see page 63.

- Tell them about the planets in the solar system.
- Tell them how hot the Sun is.
- Ask them to choose a planet that they would like to learn about.

If you don't already know about this planet, find out what it is like.

- Is it hot or cold?
- Are there any moons around the planet?
- Is it dark or sunny?
- Are there any other planets close by?

💭 Imagine you have travelled in a rocket to visit the planet.

✏️ Write a postcard from the planet.

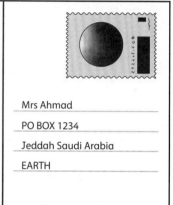

Dear Mum

I am having a great time on Mercury. During the day it is much hotter than on Earth, but at night it is much colder. I keep warm at night by wearing a space suit. There isn't much water so it is good that we brought lots with us.

Love from Mia

Mrs Ahmad
PO BOX 1234
Jeddah Saudi Arabia
EARTH

Write to the person who has helped you with this activity or choose someone else who would like to hear from you.

What we have learned about Earth's movements

See Student Book 5, pages 118–119

Home activity What I have learned…

In the word bank there are words from this module. Find them in the wordsearch. Words go across, down and diagonally.

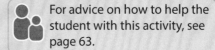
For advice on how to help the student with this activity, see page 63.

When you find each word, discuss its meaning with an adult at home.

axis	day	Earth	Moon	orbit	spin	Sun
year	revolve	rotate	movement	model	planet	star

i	r	m	z	r	o	t	a	t	e	n	c	o	c	q
u	l	o	n	t	n	v	e	v	c	o	t	w	n	d
c	k	v	h	e	f	o	l	t	a	s	y	u	v	m
y	x	e	j	a	l	o	c	t	x	k	S	x	d	o
e	g	m	a	d	v	p	s	w	i	q	r	j	r	d
a	m	e	l	e	m	g	v	y	s	v	a	p	l	e
r	o	n	r	t	p	i	h	l	a	c	c	f	v	l
r	q	t	r	t	s	p	l	a	n	e	t	t	o	f
g	b	j	z	c	d	p	t	j	i	i	r	v	r	j
E	a	r	t	h	h	e	i	t	x	n	d	k	b	k
i	d	d	v	z	t	e	t	n	d	d	a	m	i	t
e	l	p	r	s	t	a	r	k	i	f	a	w	t	z
d	M	o	o	n	c	x	g	o	k	t	p	y	i	t

6 Shadows

Extra support

Introduction

The main purpose of this module is to help students to understand how different materials produce shadows. They will investigate translucent, transparent and opaque objects and how these cast different intensities of shadows. They will explore how shadows of different sizes are produced.

The next activities allow students to explore how shadows can be useful. Students use shadows to make artwork (silhouettes) and explore how we use shadows to keep cool on sunny days. The student will invite the people at home to play with shadows by joining in a game of 'Shadow hide and seek'. They will learn that we can tell the time using shadows cast by the Sun and that sundials were used in the past. In the final activity, the student will use a solar powered calculator as a light meter.

This module will help students to practise these scientific enquiry skills:

- planning – asking questions and planning how to seek answers in a fair test (pages 80, 82, 86, 87, 88)
- predicting – stating what they think will happen and then comparing this with what actually happens (pages 80, 88)
- observation – collecting sufficient evidence, understanding the need for repeated observations or measurements (pages 79, 80, 82, 83, 84, 85, 86, 87, 88)
- recording – writing or drawing observations or stages in work and presenting results in line graphs or bar charts (pages 79, 80, 82, 83, 85, 86, 87, 88)
- making comparisons – comparing sets of evidence or data (pages 79, 80, 82, 87, 88)
- drawing conclusions – examining results to identify any patterns and/or to suggest explanations (pages 82, 87, 88)
- evaluating results – evaluating the investigation and results and suggesting ways to improve the investigation (pages 86, 87).

Ways to help

Look for shadows with the student and talk about the intensity and size of the shadows you can see. Help the student to understand that shadows change depending on the position of the light source and the source, the distances between the sources of light, the object and the surface the shadow is cast on. Give the student opportunities to work through the activities as independently as possible, providing support as necessary.

Remember to encourage and help the student but try to let them find out as much as they can on their own.

Finally, help the students complete the 'What I have learned…' summary to test their understanding and recall.

Key words

block	opaque
length	position
light	shadow
light intensity	silhouette
light source	translucent
object	transparent

Scientific enquiry words

design	plan
fair test	predict
measurements	results

 ## Helping with activities

The following guidance is intended to offer advice to the parent, or other adult at home, on how to help the student with each home learning activity.

Translucent, transparent and opaque objects (page 79)

Allow the student to safely explore some different objects at home. Encourage them to use the key words to describe materials that allow differing amounts of light through them. Take opportunities to point out the different types of materials and their uses.

Does it make a good shadow? (page 80)

This activity will reinforce the student's understanding that opaque materials block the light and cast a shadow. Allow them to explore different materials. Provide a light source such as a lamp or torch.

Making silhouettes (page 81)

This activity provides another fun opportunity to explore how shadows are cast and highlights the idea that shadows can be useful. Provide the student with some white paper for drawing the outline of the shadow and some black paper or card for them to cut out the silhouettes. Encourage different members of the family to have their silhouette made.

Using shadows (page 83)

Help the student to find some shady spots around the home and/or garden and discuss together how these places make the student feel. Then observe some less shaded or open spaces. Use the experience to reinforce that shadows can be created from trees or sunshades to keep us cool and safe from the Sun.

Shadow hide and seek at home (page 85)

Play the game, guiding the student to stand behind an object such as a tree or a wall so that the seeker cannot see the student but can see their shadow. It is best to play this game either early in the morning or late in the afternoon when the shadows will be longer.

Telling the time with shadows (page 87)

The student has explored how we can use a sundial to tell the time. They have made and tested a sundial at school, and have planned how to improve the sundial. Help the student to follow their plans to make the new version of the sundial. Find a sunny place to put the student's sundial. Help them to test the sundial's accuracy, using a clock at different times over a few days.

Measuring light (page 88)

Specialist equipment can be used to measure the intensity of light, but this activity provides an easily accessible and effective way for the student to measure light intensity at school and at home. You will need a solar cell calculator for this activity. Allow the student to explain to you how this works.

What I have learned... (page 89)

This summary activity tests the student's understanding of key terms and concepts from the module. Encourage the student to answer as many questions as they can before looking back through their Student Book for help.

Can we see through it?

See Student Book 5, pages 122–123

Class activity Does it let light through or block light?

✏️ Complete the sentences using the words in the word bank. You will need to use each word more than once.

An **object** that does not let **light** through is _____. We cannot see through _____ materials.

An object that lets a lot of light through is _____. We can see clearly through _____ materials. The glass in windows is a _____ material.

Some objects let a little light through. These are made from _____ materials.

We can see shapes on the other side of _____ materials but not very clearly.

Coloured and frosted glass are examples of _____ materials.

| translucent   opaque   transparent |

Look at the pictures. The student is looking through a different material in each one.

✏️ Label each picture with the correct word from the word bank to describe the material.

a

b

c

Can we see through it?

See Student Book 5, pages 122–123

Home learning Translucent, transparent and opaque objects

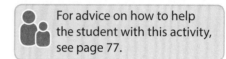

For advice on how to help the student with this activity, see page 77.

Talk to the people at home about the different types of material in the picture. Use the words 'translucent', 'transparent' and 'opaque' to describe whether they let light through.

Talk about materials that you can see in the room where you are.

Survey of materials at home

1 Find two **opaque** objects.

- Explain to the person you are working with why you know this object is made from opaque material.
- How can you prove that a material is opaque?

2 Find two **translucent** objects.

- How can you prove that the object is made from a translucent material to the people at home?

3 Find two **transparent** objects.

Draw the two transparent objects you have found.

Label your drawings to explain what the objects are used for.

Can we see through it?

See Student Book 5, pages 124–125

Home learning Does it make a good shadow?

💬 Talk to the people at home about how **shadows** are made.

For advice on how to help the student with this activity, see page 77.

- What kind of material makes the best shadow?
- Why does a transparent material not make a good shadow? What happens to the light?

✋ Which materials make a good shadow?

1. Find five objects in your home that you can test.
2. **Predict** whether each object will make a good shadow or a bad shadow.

✏️ Fill in the table with your predictions.

Object	Prediction: good shadow or bad shadow?	Result: good shadow or bad shadow?

3. Now test your prediction.
   - Use a torch or lamp as a source of light.
   - Use a white piece of paper or a white wall as a screen.
   - Shine the light on the object with the wall or screen behind it.

✏️ Did the objects cast shadows? Fill in the table with the **results**.

Did any of the objects make good shadows? _____

Did any objects surprise you? _____

Creating shadows

See Student Book 5, pages 126–127

Home learning Making silhouettes

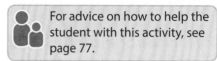

For advice on how to help the student with this activity, see page 77.

The picture shows the silhouette of the profile of a girl.

💬 Tell the people at home about **silhouettes**.

✋ Make a silhouette.

Ask someone to take part in this activity. You will make a silhouette of their profile.

You will need: a sharp pencil, a large piece of paper, a **light source** (a torch or lamp), a piece of black card and scissors.

1. Ask the person to sit about 1 metre from a wall or screen. They must sit sideways so you can make a shadow of their profile. Ask the person to sit very still.

2. Tape a piece of paper to the wall or screen. Make sure it is in the right **position** so you can see the person's shadow on the paper.

Ask permission before you tape the paper to the wall or screen.

3. Turn off the light in the room. Close the curtains or blinds. The room needs to be dark.

4. Shine a light onto the side of the person's face.

5. Draw around the shadow that is cast on the paper.

6. Stick the piece of paper onto black card. Carefully cut out the silhouette.

💬 Tell the person you have drawn that this is how Étienne de Silhouette made pictures of people in the 1700s.

Growing and shrinking shadows

See Student Book 5, pages 128–129

Class activity Investigate the size of shadows

Look at the diagrams. They show two objects with a source of light shining on them.

The shadow of each object is cast on the screen.

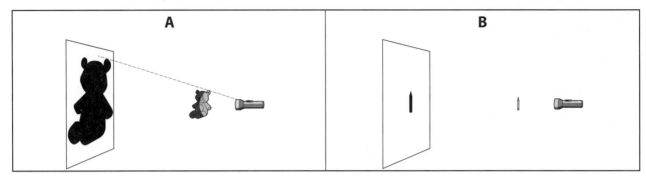

Draw a straight line from the top of the light source, across the top of the object, to the screen. One is done for you.

Draw another straight line from the bottom of the light source to the screen.

Measure the **length** of the shadow cast on the screen by the object.

Record your **measurements**.

Object A: _____ mm

Object B: _____ mm

Which object casts a bigger shadow? _____

Do bigger objects cast bigger shadows?

1 Find two objects that are similar but different sizes.

2 Measure and record the length of each object.

3 Use a light source to cast each object's shadow on a piece of paper.

4 Measure from the top to the bottom of each shadow. Record your **measurements**.

Object	Length of object (cm)	Length of shadow (cm)

Does the bigger object cast a bigger shadow?   Yes   No

Growing and shrinking shadows

See Student Book 5, pages 128–129

Home learning Using shadows

💬 Talk to the people at home about how we use shadows.

- Why do we sometimes want shadows or shady places?
- Do people deliberately make shady places?

For advice on how to help the student with this activity, see page 77.

✋ Find places that are in shadow.

Look around your home and outside to find places that are in the shade.

✏️ Where are shadows and shade useful? Draw some examples.

✏️ Where can shadows and shade be a problem? Write some ideas.

💭 Think back to your work with plants in Module 4.

Are some parts of your home too shady for plants to grow well?

What will happen to plants if you try to grow them in the shady places?

Tracking those moving shadows

See Student Book 5, pages 130–131

Class activity Shadow hide and seek

You are going to use shadows to play a game of 'Shadow hide and seek'.

Play the game in a group of four or five or with the whole class.

1 One student in your group is the seeker.

2 The seeker must close their eyes and count to 20.

3 While the seeker is counting, the other students must hide. They must stand so they cannot be seen but their shadow can be.

4 When the seeker has finished counting they can turn round.

5 The seeker must look for people's shadows. When they see a shadow, they must guess who the person is by the shape of the shadow.

6 The seeker says the name of the person they think it is. If the seeker is correct, that person is out.

7 How many people can the seeker guess correctly?

Take turns to be the seeker and the hiders.

Tracking those moving shadows

See Student Book 5, pages 130–131

Home learning Shadow hide and seek at home

💬 Teach your family how to play 'Shadow hide and seek'.

Show them how to hide behind an object and cast a shadow.

Explain that to cast a shadow they have to **block** the light.

Now play the game.

1 Choose one person to be the seeker.
2 The seeker closes their eyes and counts to 20.
3 The other people must hide.
   - Remind the hiders that they must hide so the seeker cannot see them but can see their shadow.
4 The seeker has to guess who each person is by the shape of their shadow.
   - The seeker calls out the name of the person they think it is.
   - If the seeker guesses the hider correctly, that person is out.
5 Take turns to be the seeker and the hiders.

✏️ Draw your family playing 'Shadow hide and seek'.

Don't forget to draw the shadows. How long should you draw the shadows?
What colour should you use to draw the shadows?

> For advice on how to help the student with this activity, see page 77.

Tracking those moving shadows

See Student Book 5, pages 132–133

Class activity Make a sundial

 Make and test a sundial.

> You will need: a paper plate and a pencil.

1. Carefully push the pencil through the centre of the plate. Your teacher may help with this part.
2. Write the time on the edge of the plate. For example, if it is midday write 12. Draw a straight line from the centre of the plate to the number.
3. Take the plate outside. Place it in the sunshine, on the ground or a flat surface. The pencil will cast its shadow onto your plate.
4. Turn the plate so that the shadow falls on the line you have drawn.
5. One hour later, go outside again. Draw a line where the shadow is and write the time.
6. Every hour, **predict** where you think the next line will be. Then go outside to check. Draw the line and write the time.
7. At the end of the day, **predict** where you think the line will be early the next morning.
8. Test your finished sundial throughout the next day.

Is your sundial accurate in telling the time?   Yes   No

Evaluate your sundial. Discuss some ways to improve it with other students in your class.

You are going to make an improved model at home and test it for a few days.

Step 1

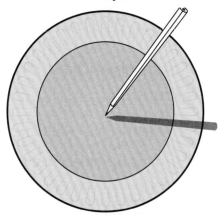

Step 4

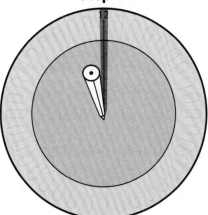

Step 6

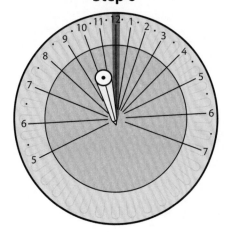

Tracking those moving shadows

See Student Book 5, pages 132–133

Home learning Telling the time with shadows

Talk to the people at home about how you made a sundial.

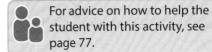
For advice on how to help the student with this activity, see page 77.

- Explain how you used shadows to tell the time.
- Tell them how you turned the sundial at the beginning of the investigation so the shadow pointed to the right time.
- Talk about the improvements you would like to make to the sundial.

 Make and test an improved sundial.

1 **Plan** how to make an improved model for a sundial.
   - What equipment will you need?
   - Will you use the same method?

Write details or draw your improved **design** in the box.

2 Test your improved sundial for a few days. Find out if it tells the time accurately.

You can record your **results** in the tables below.

Day								
_____	Time of day							
	Time on sundial							

Day								
_____	Time of day							
	Time on sundial							

3 Compare the time of day to the time shown on the sundial.

Is the sundial more accurate at some times of day than others?

_____

Did your improvements work?

_____

Measuring light intensity

See Student Book 5, pages 134–135

Home learning Measuring light

Show your family how you can use a solar powered calculator as a light meter.

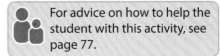 For advice on how to help the student with this activity, see page 77.

Compare light sources at home.

1 Find six sources of light at home.

2 **Plan** an investigation to compare the **light intensity** of these light sources.

Use a solar powered calculator and tissue paper in your plan.

Write your plan on a separate piece of paper.

3 **Predict** which light source is the brightest. This one will need the most sheets of tissue paper to stop the calculator display working.

I think _____ will need the most sheets of tissue paper to stop the display working.

4 Use the solar powered calculator to find out which light source is the brightest.

Record your **results** in the table. Write the number of sheets of tissue paper needed to stop the calculator display working.

Light source	Number of sheets of tissue paper	Light source	Number of sheets of tissue paper
1		4	
2		5	
3		6	

Write the light sources in order of brightness.

Least bright ⟶ Brightest

Was your prediction correct?   Yes   No

What we have learned about shadows

See Student Book 5, pages 138–139

Home activity What I have learned...

Answer the questions to help review your understanding of this module.

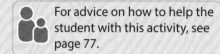
For advice on how to help the student with this activity, see page 77.

1. Circle the correct word to complete each sentence.

   a Materials that do not let light through are **transparent / translucent / opaque**.

   b Materials that let a lot of light through are **transparent / translucent / opaque**.

   c Materials that let a little light through are **transparent / translucent / opaque**.

   d **Transparent / translucent / opaque** materials make the best shadows.

2. Decide whether each statement is true or false. Circle the correct answer.

   a Smaller objects cast bigger shadows. **True   False**

   b Shadows can be useful. We can use shadows to tell the time. **True   False**

   c Solar powered calculators can be used to measure light intensity. **True   False**

3. Draw a circle around the cup that will cast the darkest shadow.

  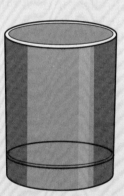

Explain your answer.

_____

_____

Quiz yourself

These quiz questions and activities are intended to encourage the student to reflect on their learning and to reinforce their developing knowledge about scientific concepts in a fun way. They are flexible enough to be individual, pair or group activities. Teachers and parents can use the quizzes in a number of ways.

- Questions can be selected from this section to supplement work carried out during each module, to act as extra tasks and support for individuals, groups and whole classes. In this way they can aid differentiation.

- Students can tackle the relevant questions at the end of each module to review learning and supplement the 'What we have learned…' sections.

- Students can undertake questions at the end of a series of modules or even at the end of the year to review learning. The questions could be set in batches over a series of lessons or even taken as a small timed test – although this is not their main purpose.

After each question the students can fill in the self-review circle to show how confident they are with that task. At Stage 5, students can give a score from 1 to 5, with 5 being very confident. If a student reports that they are not confident in a certain area, the teacher can provide remediation.

1 The Way We See Things

1 a Write the word 'source' under the sources of light.

Write the word 'reflector' under the objects that reflect light.

Write the word 'sense' under the part of the body that senses light.

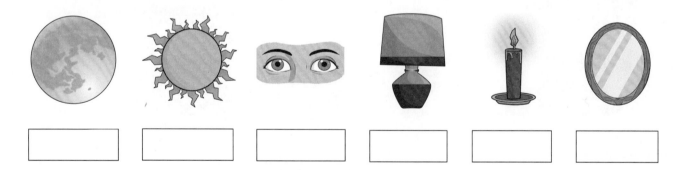

b Write down the name of another source of light and another reflector of light.

Source: _____   Reflector: _____

Self review How do I feel about this question?

2 a Complete the ray diagram. Draw a line to show the path of a beam of light leaving the lamp.

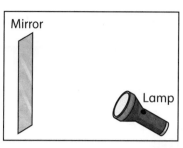

b Describe what will happen to the beam of light if a cloth covers the mirror.

_____

c Draw some mirrors inside this maze so that the person can see the flower.

How many mirrors did you need to use? _____

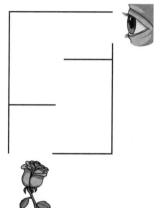

Self review How do I feel about this question?

2 Evaporation and Condensation

3 a The word bank has key words from Module 2. Can you find them in the wordsearch?

The words go across and down. Circle each word when you find it.

e	v	a	p	o	r	a	t	i	o	n	b	i
y	k	v	n	c	q	j	k	s	c	h	a	w
c	m	o	b	o	l	i	m	z	r	m	f	a
r	e	r	c	n	r	l	a	s	y	l	r	t
t	l	t	o	d	d	b	t	f	s	p	e	e
s	t	a	t	e	o	f	m	a	t	t	e	r
o	i	q	c	n	y	g	g	m	a	h	z	v
l	n	r	d	s	o	l	u	b	l	e	i	a
u	g	v	t	a	m	e	r	t	s	t	n	p
t	w	u	i	t	r	n	e	z	e	n	g	o
i	d	b	o	i	l	i	n	g	u	e	o	u
o	j	l	p	o	u	x	k	f	v	b	w	r
n	e	e	i	n	s	o	l	u	b	l	e	z

boiling
condensation
crystals
evaporation
freezing
insoluble
melting
soluble
solution
state of matter
water vapour

b Write a sentence that uses three of the words.

_____

_____

Self review How do I feel about this question?

4 a Complete the table. Write in the properties of solids, liquids and gases.

Property	Solid	Liquid	Gas
Volume			
Shape			
Density			
Ease of compression (squashing)			
Ease of flow			

b Label the particle diagrams as either solid, liquid or gas.

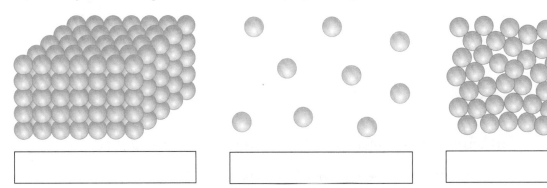

Self review How do I feel about this question?

3 The Life Cycle of a Flowering Plant

5 Draw your own flowering plant using the parts shown. Draw each part in the correct place to make a complete plant. Label all the parts of your flowering plant.

Self review How do I feel about this question?

6 Solve the clues and complete the crossword.

Across

2 This is made from the ovary and contains seeds.

4 These are often brightly coloured and scented to attract insects.

5 These creatures can move pollen from one plant to another.

8 The stage in a plant's life cycle when pollen reaches the stigma

9 This part of the plant uses sunlight energy to make sugars.

10 The first stage in the life cycle of a flowering plant

Down

1 The stage in a plant's life cycle when the seed starts to grow

3 The female part of a flower

6 The male part of a flower

7 These anchor the plant in soil and help the plant to obtain water and nutrients.

Self review How do I feel about this question?

4 Investigating Plant Growth

7 a Join the dots to find a living thing.

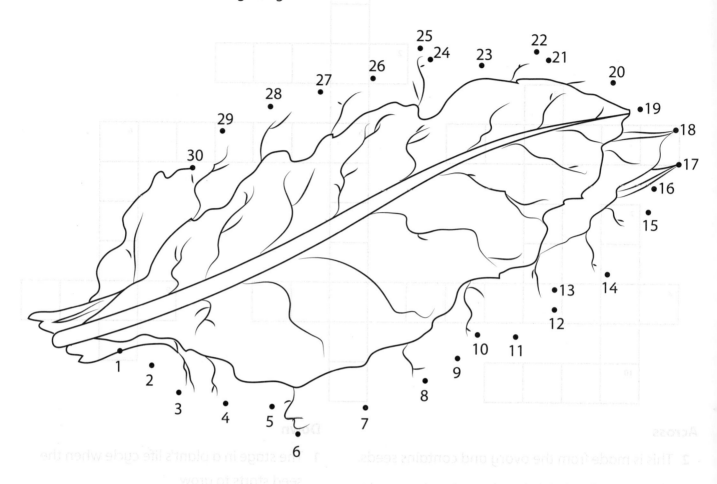

What is the name of this living thing?

_____

b Explain why the Sun is important to this living thing.

_____

_____

Self review How do I feel about this question?

8 Draw a line from each label to the correct plant.

   Use a ruler and pencil to draw your lines.

(This plant is wilting.) (This plant has had plenty of water.) (This plant has been in sunshine.) (This plant might have been in the dark.) (This plant is healthy.) (This plant has not had enough water.)

Self review How do I feel about this question?

5 Earth's Movement

9 In the daytime box draw **three** things you can only see during the day.

   In the night-time box draw **three** things you can only see at night.

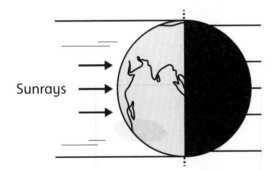

Sunrays

Day time

Night-time

Self review How do I feel about this question?

95

Quiz yourself

10 Label the planets.

Identify the missing planets and draw them in the solar system.

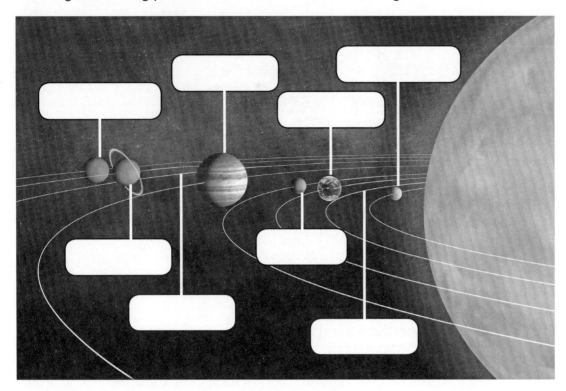

Self review How do I feel about this question?

6 Shadows

11 Draw a line from each object to its shadow.

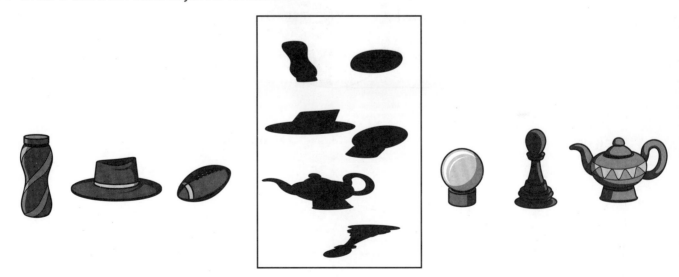

Self review How do I feel about this question?